ANAEROBIC DIGESTION OF BIOMASS

ANAEROBIC DIGESTION OF BIOMASS

Edited by

DAVID P. CHYNOWETH

Director, Bioprocess Engineering Research Laboratory, University of Florida, Gainesville, Florida, USA

and

RON ISAACSON

Manager, Renewable Resources, Gas Research Institute, Chicago, Illinois, USA

ELSEVIER APPLIED SCIENCE
LONDON and NEW YORK

ELSEVIER APPLIED SCIENCE PUBLISHERS LTD
Crown House, Linton Road, Barking, Essex IG11 8JU, England

Sole Distributor in the USA and Canada
ELSEVIER SCIENCE PUBLISHING CO., INC.
52 Vanderbilt Avenue, New York, NY 10017, USA

WITH 29 TABLES AND 41 ILLUSTRATIONS

British Library Cataloguing in Publication Data

Anaerobic digestion of biomass.
1. Biomass energy 2. Bacteria, Anaerobic
—Industrial applications
I. Chynoweth, David P. II. Isaacson, Ron
662'.6 TP360

Library of Congress Cataloging-in-Publication Data

Anaerobic digestion of biomass.

Includes bibliographies and index.
1. Biogas. 2. Anaerobic bacteria. I. Chynoweth, David P. II. Isaacson, Ron.
TP359.B48A53 1987 662'.8 87-5329

ISBN 1-85166-069-0

The selection and presentation of material and the opinions expressed are the sole responsibility of the author(s) concerned.

Typeset by Keyset Composition, Colchester
Printed in Great Britain at the University Press, Cambridge

Preface

This book is the result of an effort to document the status of anaerobic digestion processes for conversion of biomass and wastes to methane at the commercial, developmental, and research levels and to identify data gaps to commercialization and development, process limiting steps, and research needs.

The Gas Research Institute (GRI) has a program centered on the production of methane from various types of biomass and wastes. This program is devoted to the biogasification (anaerobic digestion) of these feedstocks to methane. Although anaerobic digestion has been extensively studied and applied commercially to treatment of domestic sludge, study and application of the process to other feedstocks such as municipal solid wastes and aquatic, herbaceous, and woody species has been minimal. Furthermore, the state of knowledge of this process at the fundamental level is deficient. Consequently, in certain applications, the process suffers from process instability and low conversion efficiencies. In addition, because biomass feedstocks differ in physical and chemical properties, a number of questions are raised relative to biodegradability, front-end processing, materials handling through the digester, effluent dewatering, and byproduct recovery. In view of this national effort on anaerobic digestion, there was a need to (1) document the state of the art of this process; (2) identify currently recognized and potential problems in its application for biogasification of biomass; and (3) develop a data base and documented criteria to guide research programs in this area.

The purpose of this book is to identify and evaluate the status of commercialization and process development units. In addition, it is to review basic and applied research in the areas of microbiology and process development and to provide a basis for recommending research needs and priorities related to improvement of yields, rates, process stability, net energy production, and process economics. The emphasis of the review is on biomass rather than wastes. Literature on processes for conversion of municipal, industrial and animal wastes is used only where the information is relevant to the development of processes for conversion of biomass. This review does not represent an exhaustive description of the literature, but rather draws upon recent key publications to document status and research needs in this field.

An Editorial and Advisory Panel (below) consisting of recognized experts in this field reviewed the scope, content, and R&D recommendations contained in this book. Relevant information was obtained via a literature search, personal contacts, and input from several consultants.

RON ISAACSON
DAVID P. CHYNOWETH

EDITORIAL AND ADVISORY PANEL

H. Ronald Isaacson
Committee Chairman
Gas Research Inst.
Chicago, IL 60631

David P. Chynoweth
Agricultural Engineering Dept.
University of Florida
Gainesville, Florida

Robert E. Hungate
Dept. of Bacteriology
University of California
Davis, CA 95616

Paul R. Erickson
Enviro-Energy Tech. Center
Rexnord
Milwaukee, WI 53201

Perry L. McCarty
Dept. of Civil Engineering
Stanford University
Stanford, CA 94305

Paul H. Smith
Dept. of Microbiology
University of Florida
Gainesville, FL 32611

James V. Basilico
Office of R&D
US EPA (RD-681)
Washington, DC 20460

Contents

List of Contributors

RICHARD BILJETINA
Institute of Gas Technology, 3424 South State Street, Chicago, Illinois 60616, USA

DAVE BOONE
School of Public Health, University of California, Los Angeles (UCLA), Los Angeles, California 90024, USA

Y. R. CHEN
USDA Meat and Animal Research Center, Clay Center, Nebraska 68933, USA

DAVID P. CHYNOWETH
Agricultural Engineering Department, Frazier Rogers Hall, University of Florida, Gainesville, Florida 32611, USA

KIRBY F. FANNIN
Institute of Gas Technology, 3424 South State Street, Chicago, Illinois 60616, USA

THOMAS FERGUSON
School of Medicine, University of California, Davis, California 95616, USA

H. RONALD ISAACSON
Gas Research Institute, 8600 West Bryn Mawr Avenue, Chicago, Illinois 60631, USA

DOUGLAS E. JERGER
Agricultural Engineering Department, Frazier Rogers Hall, University of Florida, Gainesville, Florida 32611, USA

ROBERT MAH
School of Public Health, University of California, Los Angeles (UCLA), Los Angeles, California 90024, USA

ROBERT SLEAT
Biotechnica Ltd, 5 Chiltern Close, Cardiff CF4 5DL, UK

RICHARD E. SPEECE
Environmental Studies Institute, Drexel University, 32nd and Chestnut Streets, Philadelphia, Pennsylvania 19104, USA

VIPUL J. SRIVASTAVA
Institute of Gas Technology, 3424 South State Street, Chicago, Illinois 60616, USA

GEORGE T. TSAO
School of Chemical Engineering, Purdue University, West Lafayette, Indiana 47907, USA

1
Overview

DAVID P. CHYNOWETH

Agricultural Engineering Department, University of Florida, Gainesville, Florida, USA

1.1 RESOURCE

Biomass (including organic residues) is an important potential source of methane when the overall strategy of gas supply for the United States is considered. Several recent systems studies have predicted that this resource can be developed to supply over 5 quads (exajoules) of energy by the year 2000 and significantly more beyond that point.[1,2] Some estimates of the potential of various categories are summarized in Table 1.1. Biomass represents a continuously renewable potential source of methane and thus offers a partial solution to the eventual prospect of fossil fuel depletion. In addition, it can be converted to methane at a variety of scales economically and thus can be tailored to supply local and regional gas needs as well as nationwide demands. Large-scale use of biomass as an energy resource envisions not only use of portions of existing croplands, forests, and collected wastes as feedstocks for

TABLE 1.1
Potential US Biomass: Energy in the Next Century* (Quads/year)

Crop	14
Forestland	23
Marine biomass	14
Wastes	4
Total	55

*Adapted from Ref. 2; numbers represent the gross energy value of the biomass.

conversion processes, but also commitment of additional areas of land and water for energy farms and increased collection of waste feedstocks such as agricultural and forest residues. The use of biomass as an energy resource would represent a short-term, aboveground carbon cycle, thereby minimizing alteration of atmospheric carbon dioxide concentrations. Organic waste conversion to useful energy forms has the added benefit of solving environmental problems associated with alternative methods of disposal.

1.2 CONVERSION PROCESSES

Processes for conversion of biomass to methane may be classified into two categories, thermal and biological. Economic application of thermal processes is limited to feeds with either a low water content or having the potential to be mechanically dewatered. This limitation is linked to energy requirements related to evaporation of water in order to achieve the high temperatures required for the process. Thermal processes also produce a mixture of gaseous products that must be upgraded to methane and are only economic at larger scales. The major advantage of thermal processes is their ability to effect total conversion of organic matter at rapid rates.

This report is concerned with biological gasification, better known as anaerobic digestion, which is a low-temperature process that can process wet or dry feeds (with added water) economically at a variety of scales. The product gas is composed primarily of methane and carbon dioxide with traces of hydrogen sulfide. The major limitation of this process is that conversion is usually not complete, often leaving as much as 50% or more of organic material unconverted. The process rates are significantly lower than those of thermal processes and bacteria involved require a balanced diet of nutrients that may not be available in a feedstock. Selection of a biological or thermal gasification process for the production of methane from biomass, therefore, depends on a number of factors including feed moisture content, feed biodegradability and nutrient content, and size of the facility.

1.3 MICROBIAL ECOLOGY OF METHANOGENESIS

Microbial methanogenesis is a natural process occurring in anaerobic environments such as ocean and lake sediments and animal digestive

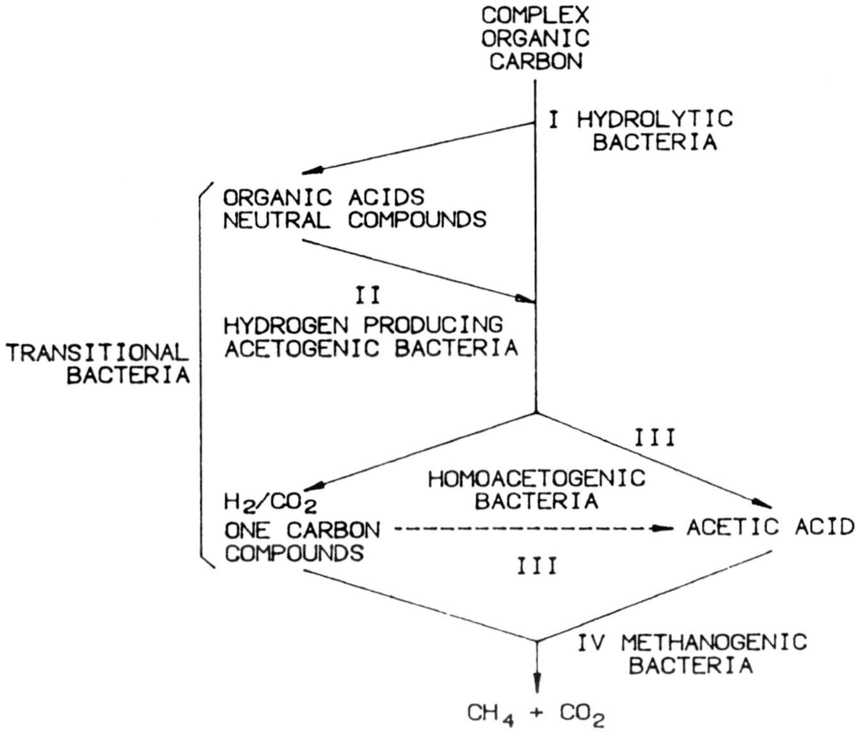

FIG. 1.1. Categories of metabolically distinct bacteria in the methane fermentation.

tracts. Man is actively engaged in harnessing and optimizing this process for decomposition of wastes and biomass into useful fuels. Methanogenic decomposition, as illustrated in Fig. 1.1, is a process occurring only under strict anaerobic conditions where mixed populations of bacteria decompose organic matter to carbon dioxide and methane. Methanogenesis has traditionally been considered a two-step process involving the acid-forming bacteria which hydrolyze and ferment organic compounds (that is, carbohydrates, lipids, and proteins) to fermentation products such as organic acids, alcohols, neutral compounds, hydrogen, and carbon dioxide, and methane-forming bacteria that convert these fermentation products to methane and carbon dioxide. As knowledge of methanogenic bacteria has accumulated, it has become apparent that their substrates were restricted primarily to acetate, hydrogen, and carbon dioxide, and a separate group of bacteria, the hydrogen-producing acetogenic bacteria, convert hydrolysis and fermentation products to acetate, hydrogen, and carbon dioxide. In many cases, hydrogen and acetate have been shown to inhibit the acetogenic bacteria. Consequently, the balanced fermentation is dependent upon constant removal of these products by the methanogenic bacteria. More recently, a fourth group of bacteria has been identified in the fermentation, the homoacetogenic bacteria, which ferments a wide spectrum of compounds such as hydrogen, carbon dioxide, and formate to acetate. The significance of this group in the overall fermentation is currently unknown. The first group responsible for hydrolysis is now referred to as the hydrolytic bacteria. In this review the bacteria are discussed in three groups: (1) hydrolytic bacteria, (2) transitional bacteria (those that convert soluble organic matter into methanogenic substrates), and (3) methanogenic bacteria.

During the past 10 years, literature on the microbiology and biochemistry of methanogenic bacteria has increased exponentially with isolation of several new species, development of more rapid techniques for study, and increased interest in this unique group of organisms. Beginning with the discovery of interspecies hydrogen transfer in 1967,[3] several species of hydrogen-producing acetogenic bacteria have been isolated. Although the overall niches of these microbial groups are beginning to unveil, it is important to realize that knowledge of the physiology, metabolism, and genetics of these organisms is in its infancy. Hydrolytic bacteria are the least studied of organisms in the methane fermentation even though hydrolysis is apparently the overall rate-limiting step in many of the methanogenic environments. With the exception of the rumen, knowledge of the microbial ecology of methanogenesis in anaerobic environments, including anaerobic digesters, is superficial. This deficiency in knowledge may be attributed to several factors:

(1) The study of obligate anaerobic bacteria is tedious and time-consuming and requires a large manpower and time effort for isolation of important organisms and study of their ecological relationships.
(2) The methane fermentation in a particular environment involves numerous species of bacteria operating in a consortium and involving numerous interspecies activities.
(3) Each anaerobic environment may differ in the types of bacteria involved in methanogenesis, depending on differing factors such as substrate, retention time, temperature, pH, and fluctuations in

environmental parameters. Although certain general properties are common from one environment to another, each environment may have its own unique population of bacteria and associated microbial activities.

(4) Obligate interdependencies of bacteria involved in the methane fermentation make it difficult to isolate certain bacterial participants into pure culture. Furthermore, pure cultures exhibit different properties under pure culture conditions than in mixed culture.

Direct benefits of basic research on the microbiology of such environments are limited. Rather, they add to the pool of knowledge on the ecology of the system which ultimately can be applied to the understanding and improvement of the process. Continued emphasis on the microbiology and biochemistry of anaerobic digestion will provide information that can be utilized to understand factors limiting process yields and rates and affecting process control and stability. For example, the microbial ecology of the rumen is reasonably well known and is stimulating investigation of rational approaches for manipulating the microbial community to achieve more effective production of milk and meat.

1.4 OVERVIEW OF ANAEROBIC DIGESTION PROCESS

A generalized scheme for the anaerobic digestion process based on methane fermentation is shown in Fig. 1.2. Upon harvest or collection, biomass or wastes are chopped and ground for size reduction, and possibly subjected to some kind of pretreatment to enhance biodegradation. In a continuously stirred tank reactor (CSTR) the feed is either added directly or as a slurry at a typical loading rate of 0·1 lb volatile solids per ft^3-day (1·6 kg/m^3-day) and a hydraulic retention time of 15 to 20 days. In this continuously mixed reactor, which is usually operated at a mesophilic temperature of 35 °C, a mixed population of bacteria effect conversion of about 50% of the organic matter (volatile solids) to methane and carbon dioxide, at a mole percent ratio of 60:40. This is equivalent to a methane yield of 4 SCF/lb (0·25 m^3/kg) volatile solids (VS) added and a methane production rate of 0·4 vol/vol culture-day. Solids in the effluent may be settled in an anaerobic secondary digester and recycled (directly or following posttreatment), processed as fertilizer

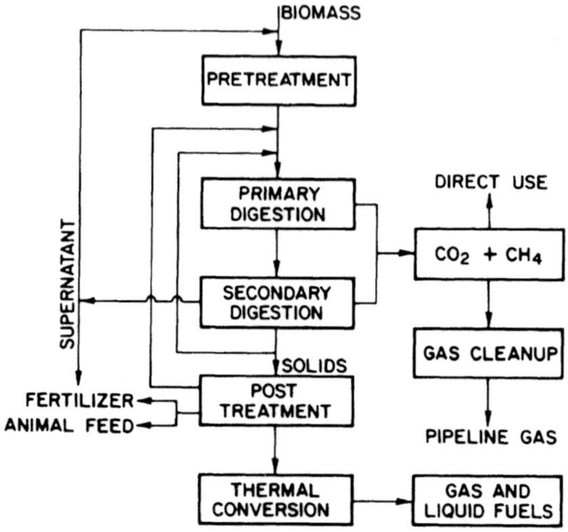

FIG. 1.2. Generalized anaerobic digestion process scheme.

or animal feed, or subjected to thermal conversion. Supernatant may be recycled, used as fertilizer, animal feed supplement, or processed into a form suitable for disposal. Product gas can be utilized directly or treated to remove carbon dioxide and traces of hydrogen sulfide.

The above refers to a conventional digester that might be found in a waste treatment plant for treatment of sludge; a number of variations in this scheme are possible. Methane yields from this process are limited by the fact that only a fraction of the organic matter in the feed can be hydrolyzed under anaerobic conditions. Process rates are limited by the slow growth rates of methanogenic bacteria for soluble feed constituents and hydrolysis rates for particulate feed components. Research to overcome these limitations is therefore focusing on methods of increasing solids and microorganism retention in the reactor and pretreatment to improve biodegradability.

1.5 PROCESS EVALUATION

The major criteria used to evaluate performance of anaerobic digestion of different feeds or new digester reactor designs are methane yield, methane production rate, reduction in organic matter, culture stability, thermal efficiency, and process economics. Methane yield is the quantity of methane produced relative to the quantity of organic matter added, and is usually reported as SCF/lb (m^3/kg) VS (ash-free dry weight) added. Methane production rate refers to the quantity of methane generated in volumes of methane per volume of active reactor per day. Organic reduction efficiency is an important performance parameter because it directly influences methane yield and is inversely related to the quantity of process residues requiring further processing or disposal. This parameter is determined by a process materials balance in terms of organic matter reduction. Certain combinations of feedstocks and operating conditions may result in microbial population imbalance and related reduction or failure of digester performance. This condition is characterized by reduced gas production, increased volatile acids concentration, and reduced pH. Finally, process evaluation must consider energy balances and economics relative to the overall biomass production or waste collection and treatment system. These criteria and the factors influencing them are outlined briefly below.

1.6 FACTORS LIMITING YIELDS

In any anaerobic digestion process that is not inhibited or kinetically limited, two major factors affecting methane yields are feedstock composition and inoculum characteristics. The composition of the biodegradable organic compounds can, for example, influence methane yield in that reduced compounds such as fats and protein produce a higher percentage of methane than oxidized compounds such as sugars. Ultimate methane yields are, however, influenced principally by the biodegradability of the organic components. Certain natural organic compounds such as lignin, for example, are refractory to decomposition under anaerobic conditions even at long residence times. This explains why organic matter accumulates in lake sediment and as fossil fuels such as peat, oil, and coal. This refractory property is related to the lack of enzymes in the anaerobic bacteria and an oxygen requirement by many enzymes to carry out the initial hydrolytic reactions. It has also been established that certain compounds can complex with others rendering them resistant to anaerobic decomposition. For example, lignin affects decomposition of cellulose in this manner. A variety of physical, chemical, and biological treatment techniques are under consideration for improvement of biodegradability. Because most of these techniques were developed for improvement of animal digestibility or conversion of biomass to ethanol, research on application of these methods for anaerobic digestion has been minimal.

Lack of decomposition of certain components of feed substances may be attributed to the absence of organisms capable of their degradation (rather than the inherent refractory nature of the compounds). It is generally thought that for most types of organic wastes and biomass, a viable inoculum can be developed from any environment in which anaerobic methanogenic decomposition of a mixture of organic compounds is occurring naturally; for example, anaerobic sewage digesters, anaerobic lake sediments, or animal feces. However, there is disagreement on this point; it would be expected that an inoculum adapted for a long period of time on a substrate would give improved performance over an unadapted inoculum. Microbial population cycling may also occur and influence performance even under continuous feeding conditions.

1.7 FACTORS LIMITING KINETICS

Process kinetics reflected as methane production rate are important to design and operation of anaerobic digesters since rapid rates permit higher loadings or smaller reactor size. A number of mathematical models have been developed to predict kinetics of various digester designs. Most of these are based on the Monod model for continuous culture of bacteria. Representative methane production rates for various feeds and operating conditions range typically from 0·5 to 5 vol/vol of active reactor per day for particulate feeds to over 10 vol/vol-day for soluble feeds. In general, higher rates result in reduced methane yields. Rates of most CSTR digesters with low hydraulic retention times are limited by the activity of methanogenic bacteria. Adjustment of design or operation to remove this limitation may result in a hydrolysis-limited system. Major factors affecting the kinetics of the anaerobic digestion process are feed concentration, temperature, nutrients, mixing, particle size, toxic feed components, and feeding frequency. Each of these will be discussed in greater detail in this report.

1.8 DIGESTER DESIGN

Development and design of new and innovative digestion schemes should have the goals of maximizing methane yields and optimizing production rates, increasing process stability, decreasing processing energy requirements, and decreasing digester costs through simplification of design and operation. From the previous discussion it can be surmised that optimization of yields can be best achieved through selection of biodegradable feeds, pretreatment, and development and optimization of inoculum. Optimum process kinetics without sacrifice in yields may be achieved by schemes with long retention of microorganisms and solids, thermophilic temperature, minimum particle size, nonlimiting nutrient levels, continuous feeding, and absence of toxic feed components.

Optimization of anaerobic digestion for high yields and rates may not be consistent with obtaining the highest net energy output and best economics. For example, higher temperatures, particle size reduction, and heat pretreatment are all factors that may increase yields and rates, but may result in overall reduction in net output process energy and increased process cost.

It is becoming apparent that in many applications the conventional stirred tank reactor is not the optimum configuration for anaerobic digestion. Limitations of this reactor include washout of unreacted solids and active microorganisms at higher loadings, difficulty in achieving complete mixing, energy requirements associated with mixing, and disruption of microbial consortia by mixing. Several innovative digester configurations are under development with the objectives of increasing methane yields and production rates, increasing process stability and net energy output, and simplifying design and operation and improving economics. The basic principles of several of these designs (illustrated in Fig. 1.3) and conventional batch and stirred tank reactor digestion are described in the following sections to provide background for their reference and discussion in this report.

1.8.1 Batch

The simplest and most common type of digestion used worldwide is batch. Feed is added to an inoculum derived from a previous digester or other source and the fermentation is allowed to proceed until gas production ceases or becomes negligible. This type of operation is used widely at the household or farm scale. Large-scale applications of batch digestion in the form of landfills have been used for municipal solid waste and have been considered for biomass. Recent studies have demonstrated the advantages of operating batch digesters with high solids concentrations. A major disadvantage of batch digestion is that it is relatively unstable and uncontrollable due to changes in the bacterial population during the course of the fermentation. These changes can lead to

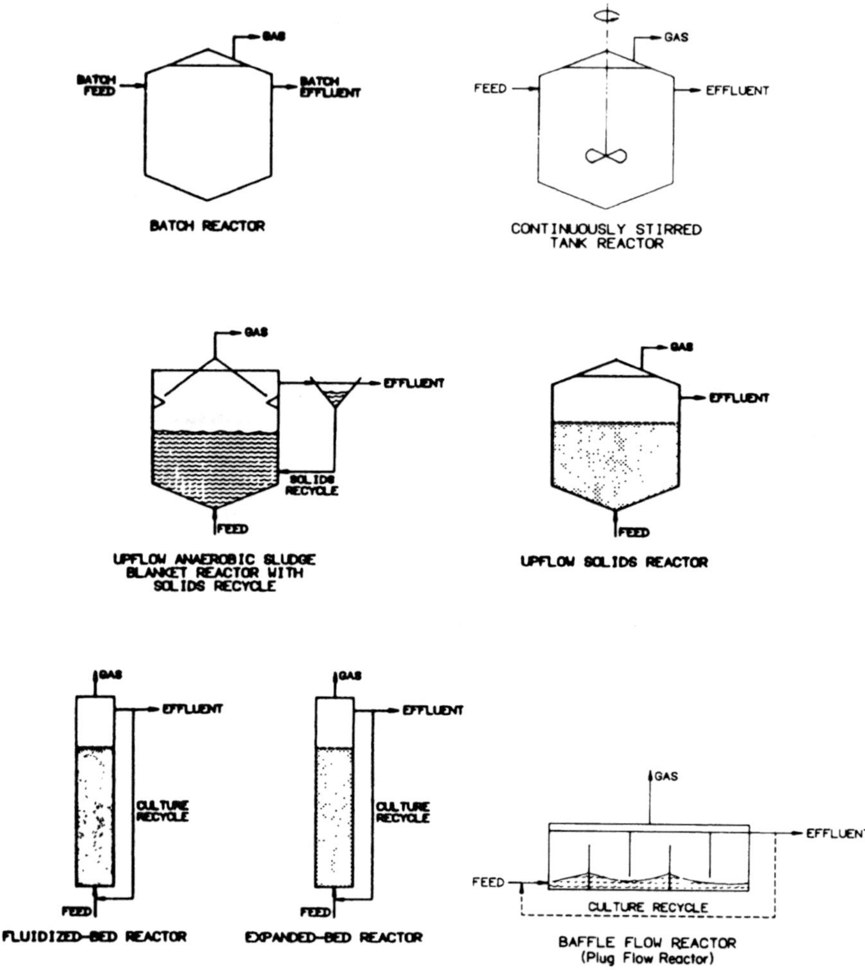

FIG. 1.3. Biological gasification process schemes.

population imbalance, digester failure, and variations in the quantity and composition of product gas.

1.8.2 Continuously Stirred Tank Reactor (CSTR)

The CSTR represents the conventional digester typically used in waste-water treatment plants. Feed is added and an equivalent volume of effluent is removed on a continuous or intermittent basis from a con-

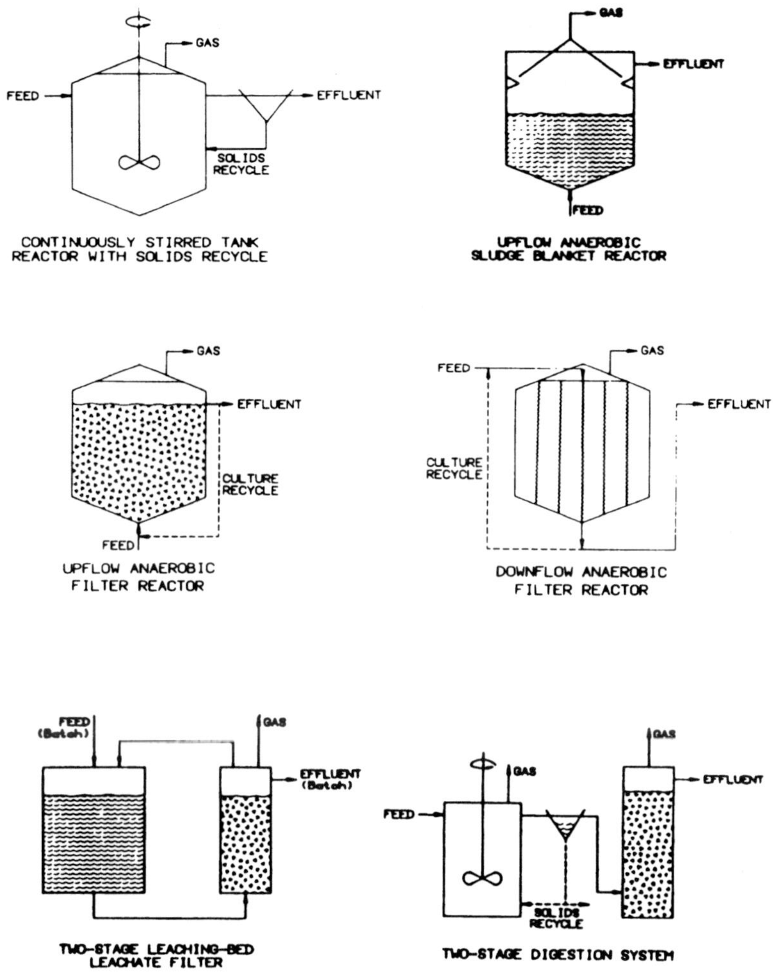

FIG. 1.3.—*contd.*

tinuously mixed reactor. In this reactor, the solids and liquid retention times are equal. Thus, as the loading is increased the retention time decreases and unreacted solids and microorganisms wash out. The result is reduction in the initial hydrolysis reactions and washout of slow-growing bacteria. The latter may lead to imbalance in the bacterial population, accumulation of volatile acids, and digester failure. Another limitation is that complete mixing is difficult to achieve in large reactors.

1.8.3 Continuously Stirred Tank Reactor With Solids Recycle (CSTR/SR)

This scheme results in improved kinetics through increasing the retention of microorganisms and unreacted solids by settling and recycling solids. This process has been widely applied for treatment of soluble industrial wastes and other feeds where solids recycle is necessary to maintain a large population of active bacteria in the anaerobic digester. More recent application of this concept to particulate feeds has also proven successful. One limitation of this design, however, is that extra equipment is required to actively increase the solids and microorganism residence time.

1.8.4 Plug-Flow

A plug-flow reactor applied to anaerobic digestion is an oblong, unmixed reactor that receives feed at one end with effluent removal at the opposite end. While some vertical mixing of the digester contents occurs during gas production, longitudinal mixing is minimal, thus promoting phase separation of the digestion process. Because of the absence of mechanical mixing, solids are passively settled and have longer retention times than liquids in this system. Feedstock hydrolysis and acid production occur initially in the digester while methane production increases toward the effluent end. However, microorganism washout can occur in this system unless an inoculum is provided with the feed or solids recycle is incorporated into the operation. Another method that has been successfully applied for reducing microorganism washout in plug-flow reactors is the installation of vertical baffles that create dead space to harbor active organisms. This modification, referred to as a baffle flow reactor, has advantages that include improved stability, higher conversion efficiencies, simplicity, and low cost.

1.8.5 Upflow Anaerobic Sludge Blanket

The upflow anaerobic sludge blanket (UASB) digester concept is based on the upward movement of soluble organic feeds through an expanded dense blanket of solids consisting primarily of microorganisms. As the liquid moves through the solids, increased solids and microbial residence times are achieved in a single digester. Various types of separators are employed to separate gas from solids and to prevent solids flotation and washout. High methane production rates and yields have been achieved with this system using soluble feeds.

A modification of this process, known as the upflow solids reactor, has been applied to particulate feeds such as kelp and water hyacinth. In the upflow solids reactor, unreacted solids and microorganisms are maintained in the reactor by passive settling. The solids are not expanded as in the UASB reactor. Significantly higher methane production yields and rates have been achieved than in the CSTR.

1.8.6 Attached Film

Attached film digesters such as anaerobic filters and expanded- or fluidized-bed reactors contain solid media upon which bacteria attach. This prevents their washout with continuous feeding and effluent removal while promoting stable intermicrobial associations. The anaerobic filter consists of a filter bed filled with inert support material such as gravel, rocks, charcoal, or plastic media. The media may be random or oriented and the flow up or down. This type of system results in rapid-rate biomethanogenic conversion of soluble wastes. The anaerobic filter is also under consideration for the methane-phase component of two-phase digestion. Feeds with high suspended solids levels are not suitable, however, as they rapidly clog the system.

A second type of attached film digester contains inert particles small enough to be hydraulically expanded or fluidized. Bacteria attach to particles and are retained in the reactor. There are two major advantages over packed-bed systems: (1) a larger surface area is available for attachment of organisms, and (2) particulate feed can pass through the inert particles without removing them from the reactor. This type of reactor can be used as a combined-phase digester for particulate and soluble feeds or as a methane-phase for two-phase systems.

1.8.7 Two-Phase

As discussed above, anaerobic digestion is carried out by two groups of bacteria (methanogenic and nonmethanogenic) that differ significantly with respect to physiology, nutritional requirements, growth and metabolic characteristics, environmental optima, and sensitivity to environmental stress. In conventional digestion, it is impossible to optimize conditions for both groups of bacteria. Theoretically, physical separation of the nonmethanogenic and methanogenic phases permits: (a) maintenance of appropriate densities of acid- and methane-forming bacteria; (b) maximization of rates through independent control of operation variables such as temperature, pH, oxidation–reduction potential, biomass recycle, retention time in each phase; and (c) enrichment of the methane content of the product gas, since most of the diluting carbon dioxide is produced in the nonmethanogenic first phase. Phase separation should reduce instability of performance caused by fluctuation in feedstock loading, pH, and toxic feed components. Reactors, such as plug-flow and upflow solids digesters, result in phase separation in a single reactor. Other two-phase schemes promote separation of nonmethanogenic and methanogenic bacteria in different reactors. The benefits of two-phase digestion of particulate feeds are not yet well documented by actual performance data.

1.9 SUMMARY

It is apparent that no single reactor design or set of operating conditions is suitable for all applications. Major factors influencing design and operating conditions include:

Chemical characteristics of feed
Concentration of feed biodegradable matter
Concentration of feed particulate solids
Density of raw and digested feed
Scale of application
Continuity of feed availability
Desired product (methane yield or VS reduction).

REFERENCES

1. *Energy From Biological Processes. Volume I.* Congress of the United States, Office of Technology Assessment, Washington, DC, OTA-E-124, July 1980.
2. Klass, D. L., 'Energy From Biomass and Wastes: 1981 Update,' in *Symposium Papers: Energy From Biomass and Wastes VI*, 1–61, Chicago, Institute of Gas Technology, 1982.
3. Bryant, M. P., Wolin, E. A., Wolin, M. J. and Wolfe, R. S., '*Methanobacillus omelianskii*, a Symbiotic Association of Two Species of Bacteria,' *Arch. Mikrobiol.* **59**:20–31, 1967.

2
Hydrolytic Bacteria

ROBERT SLEAT* and ROBERT MAH

School of Public Health, University of California, Los Angeles, California, USA

2.1 INTRODUCTION

The bioconversion of organic matter to methane is a complex anaerobic fermentation involving the concerted action of many microorganisms with a wide variety of metabolic characteristics. In many cases large, insoluble polymers make up a major portion of biomass; consequently, the hydrolysis of these high-molecular-weight, polymeric compounds into their component units is the necessary first step in anaerobic digestion. Only after hydrolysis takes place can the acid-forming bacteria ferment the subunits to a group of extracellular intermediates including volatile organic acids, H_2, and CO_2. These products eventually give rise to the methanogenic precursors H_2, CO_2, and acetate, which are converted to CH_4 by the methanogenic bacteria.

Hydrolytic reactions generally limit the amount of methane produced during anaerobic digestion of biomass. The concentrations of soluble, extracellular intermediates in anaerobic digesters are usually small in proportion to their production and degradation rates; volatile organic acids such as acetic, propionic, and butyric acids are intermediates which turn over about once per hour or less.[1] The loss of such compounds in the effluent is not significant to the methane-production potential because of their low concentrations and high turnover rates.[1] Furthermore, an increased concentration of these intermediates stimulates their conversion to methane and yet may not significantly inhibit hydrolytic reactions.[1,2]

Most of our knowledge of the anaerobic microbial breakdown of biomass polymers comes from studies of the microbial ecology of the rumen.[5,6] Thus a great deal of our conceptual understanding of biomass decomposition in anaerobic digesters is obtained by extrapolations of our

*Present address: Biotechnica Ltd, 5 Chiltern Close, Cardiff CF4 5DL, UK

understanding of the rumen microbial ecosystem. This chapter will review our knowledge of the anaerobic hydrolysis of the major polymeric biomass compounds — homopolysaccharides (cellulose and starch), heteropolysaccharides (hemicelluloses), pectins, lignins, proteins, and lipids.

2.2 CELLULOSE

Interest in cellulose as an abundant fermentable substrate for the production of ethanol and chemical feedstocks is reflected by the number of recent reviews on this subject.[7,8] Cellulose, the single most abundant biopolymer in nature, comprises 45% to 98% of the dry weight of wood, jute, flax and cotton.[9] In the cell walls of higher plants it is generally associated with both lignin and hemicelluloses as complex lignocelluloses. Cellulose is a long linear polymer of D-glucose residues joined by β-D-1,4-linkages. Since each of the three hydroxyl groups of each glucosyl unit may participate in hydrogen bonding, the cellulose fiber has remarkable strength. The formation of fibers may also sterically limit the availability of cellulose to the action of cellulase.

Cellulose is readily degraded aerobically by both bacteria and fungi; the fungal cellulases from aerobic systems are the most extensively characterized, but less is known of the aerobic bacterial cellulases.[9,11] Aerobic conditions and organisms are not present in anaerobic digesters where cellulose is degraded by anaerobic organisms.

The cellulolytic organisms which predominate in anaerobic digesters are all anaerobic and sensitive to oxygen; they were not well known until Hungate developed roll-tube techniques.[12] These culture techniques were used to isolate a number of anaerobic, cellulolytic rumen bacteria, and Hungate isolated the first important, actively cellulolytic rumen organism, *Bacteroides succinogenes.*[13] Other strains of this Gram-negative rod were subsequently isolated[14–16] and their physiology was studied in detail.

Later cellulolytic isolates were predominantly non-spore-forming Gram-negative rods and Gram-positive cocci. Another abundant cellulolytic, Gram-negative rod, *Butyrivibrio fibrisolvens*,[17,18] is less active. *Ruminococcus flavefaciens*[12] and *Ruminococcus albus*[19] are the predominant cocci. Gram-positive, spore-forming, cellulolytic rods are probably less important cellulose degraders in the rumen; examples are *Clostridium lochheadii*,[19] *Clostridium longisporum*,[19] and *Clostridium cellobioparus*[20] (*Clostridium cellobioparum*[21]). Other, less abundant species such as *Eubacterium cellulosolvens*[22] and thermophilic clostridia[23,24] have also been isolated.

A large number of anaerobic ciliated rumen protozoa[5,6] contribute significantly to cellulose fermentation. The flagellated rumen protozoa, which are the major cellulolytic organisms in the termite gut, are not important cellulose hydrolyzers in the rumen and have not been found in anaerobic digesters. In fact, some flagellated rumen protozoa are actually stages in the life cycle of phycomycetous fungi. One of these, *Neocallimastix frontalis*, is actively cellulolytic.[25–27]

Cellulolytic bacteria in some anaerobic digesters are predominantly non-spore-forming rods, and these may be similar to strains from the rumen.[12,28] The sum of cellulose- and hemicellulose-hydrolyzing bacteria in swine manure digesters is less than 0·1% of the total bacterial population.[29] Characterizations of cellulolytic bacteria from digesters are not common, but recent papers have described one highly cellulolytic vibrioid bacterium from digester sludge[30] and two spore-forming cellulolytic bacteria from a woody-biomass digester.[3,4] The occurrence and role of anaerobic cellulolytic fungi or protozoa in digesters are even less clear. Cellulolytic bacteria may potentially enter the municipal waste digester from runoff water or from the feces of man and animals (including ruminants). These bacteria are known to be present in the human colon,[27] rat cecum,[32] horse large intestine,[33] guinea pig cecum,[34] soil,[35] and estuarine sediments.[36] Cellulases may be cell-associated (as in *Bacteroides succinogenes*[37,38]) or extracellular (as in *Clostridium thermocellum*[39–41] and *Ruminococcus flavefaciens*[42,43]). The extracellular cellulases are better studied, although much remains to be learned about their complex nature and modes of action. *Clostridium thermocellum* contains both endoglucanase (β-1,4-glucan glucanohydrolase) and exogluconase (β-1,4-glucan cellobiohydrolase) activities.[40] This cellulase is not oxygen labile and is thermally stable; it adsorbs strongly to cellulose. Calcium ions and a thiol reducing agent stimulate its activity.[39] The pH optimum of the exoglucanase is 5·6[40] to 5·7;[39] the endoglucanase has a pH optimum of 5·2[40] or 6·1.[39] *Clostridium thermocellum* cellulase lacks β-glucosidase or cellobiase activity.[39,40] The cellulase is not constitutive, and its production may be induced by short chain cellulose molecules. The endoglucanase of the cellulase complex acts randomly over the linear cellulose molecule, and the end result of the combined action of the exoglucanase and the endoglucanase is the production of cellodextrins that are readily transported into the cell for further catabolism. In *Ruminococcus albus* the cellulase complex is active between pH 6·0 and 6·8, and its temperature optimum is 45 °C.[44] These data and the partial inhibition of cellulase by oxygen suggest that multiple enzymes are produced. In contrast to the inducible enzymes of *Clostridium thermocellum*, the cellulases of *Ruminococcus flavefaciens* and *Acetivibrio cellulolyticus* are probably constitutive; cellobiose does not suppress their formation. The cellulase complex in both organisms contains both exo- and endo-1,4-β-glucanase activity.[43,45] In *Ruminococcus flavefaciens* exoglucanase activity is predominant and a cellulose swelling factor may enhance cellulolytic activity. Cellobiase activity in *Acetivibrio cellulolyticus* is cell associated.[45] The degradation of native cellulose requires both exo- and endoglucanase activities. Crystalline cellulose is highly resistant to endoglucanase activity. The progressive action of exoglucanase on the non-reducing ends of highly ordered cellulose molecules results in the gradual breakdown of the crystalline cellulose array which is made available for endoglucanase activity.

2.3 HEMICELLULOSE

Hemicellulose is a short, branched-chain polysaccharide which is a major plant constituent, comprising as much as 40% of the biomass.[46] This polymer varies considerably in structure and may be composed of hexoses, pentoses, uronic acids, and other sugars. Most hemicelluloses are β-1,4 linked; however, galactose-containing hemicelluloses are β-1,3 linked.[47] The most abundant hemicellulose, D-xylan, has a D-xylose backbone with L-arabinose as a side chain.[46] Other types of hemicelluloses are the D-mannans, D-glucose-D-mannans, D-galacto-D-gluco-D-mannans and L-arabino-D-galactans.[47]

Aerobically, hemicelluloses are readily degraded by a variety of fungi and bacteria. The properties and mode of action of some of these aerobic hemicellulases were reviewed by Dekker and Richards.[48] The predominant hemicellulose degraders found in the rumen are Gram-negative, non-spore-forming rods and cocci. *Bacteroides ruminicola*[15,49] and *Bacteroides fibrisolvens*[15,17] may each use D-xylan as a carbon and energy source. *Ruminococcus flavefaciens* and *Ruminococcus albus* may also degrade hemicellulose, but they are incapable of utilizing isolated hemicelluloses as energy sources.[15,50] The rumen spirochetes also catabolize hemicellulose; they metabolize both xylan and arabinogalactan.[51] Lastly, the rumen fungus *Neocallimastix frontalis* grows by using xylan.[27] Other habitats have also yielded axenic cultures of hemicellulolytic organisms.

Bacteroides ruminicola was isolated from a piggery waste digester[52] and other species of *Bacteroides* were isolated from the human colon. These latter bacteria fermented a variety of plant polysaccharides including arabinogalactan.[53] In addition, *Bacteroides adolescentis*, *Bacteroides infantis* and *Peptostreptococcus productus* all degrade xylan.[54]

The varied structure and composition of hemicellulose is reflected in the range of hemicellulases. Dekker and Richards[48] reviewed the hemicellulases produced by the rumen microbial population. L-Arabinases, D-galactanases, D-mannanases and D-xylanases are all detectable in the culture supernatant fluid of mixed rumen microbial populations. L-Arabinase from the rumen can liberate L-arabinose from arabinoxylan and is an α-1,3-L-arabinofuranosidase. D-Galactanase and D-mannanase degrade D-galactan and D-mannan randomly and are classified as endo β-1,4-galactan galactanhydrolase and endo β-1,4-mannan mannanhydrolase, respectively. The D-xylanases are also of the endo type and have a high affinity for the linear D-xylans. The presence of substituent sugars on the xylan backbone inhibits the action of the D-xylanases. Thus, efficient degradation of hemicellulose requires the action of multiple enzymes working synergistically. Removal of the side group substituents from the backbone chain is accomplished by specific enzymes; this allows the randomly acting endohydrolases to cleave efficiently the linear backbone polymer into small oligosaccharides. By the action of enzymes such as β-D-xylosidase (β-D-xyloside xylohydrolase), these can then be further hydrolyzed to D-xylose, which can be taken up and metabolized. Endoxylanases are generally extracellular enzymes, although in *Bacteroides ovatus* and *Bacteroides eggerthii* these may be cell bound.[55]

2.4 LIGNIN

Lignin comprises 20–30% of the dry weight of vascular plant tissues[56] and plays a role in cementing the polysaccharide component into plant cell walls, both physically and chemically. Lignin is a complex, three-dimensional, aromatic polymer formed by the nonenzymatic, random coupling of phenylpropane-based free radicals. The major linkage in lignin is the β-biphenyl ether bond. Lignin is generally considered resistant to anaerobic microbial degradation.[8] The anaerobic biodegradability of plant components such as cellulose and hemicellulose may be significantly influenced by their lignin content. Demonstration of direct anaerobic lignin degradation was tested with [^{14}C]-synthetic lignins[57] of aryl-labeled coniferyl alcohols. The radioactive substrate was incubated with stacked steer-manure and freshwater lake sediment in the presence and absence of nitrate. No $^{14}CO_2$ was detected under such conditions. Aerobic incubation gave rise to $^{14}CO_2$ production. At high temperature and pH, natural lignin is partially solubilized, and low-molecular-weight aromatic compounds (1400, 700, and 300 daltons) are chemically produced under these conditions. Anaerobic degradation of these synthetically solubilized lignin compounds may be accomplished by anaerobic, mesophilic digestor sludge.[58] In these experiments, after 30 days' incubation the molecular weight profiles indicated a degradation of material in the highest (1400) original molecular weight fraction. This was attributed to cleavage of the intermonomer ether linkages to form single ring and smaller compounds. These latter compounds are readily digested to methane and carbon dioxide.[59–61] Similar experiments with synthetically produced lignin demonstrated that only the lowest-molecular-weight compounds (<300 daltons) were anaerobically degraded when incubated with freshwater lake sediments.[62] Nevertheless, it is clear that not all ether linkages are totally refractory to anaerobic degradation. Anaerobic lignin-degrading microorganisms remain to be characterized. However, two types of anaerobic bacteria capable of dissimilating simple aromatic compounds were recently cultured.[63,64] The first, a motile, Gram-negative, rod-shaped bacterium, degraded benzoate in coculture with an H_2-utilizing organism.[63] It produced acetate and presumably CO_2 and H_2 or formate from benzoate. The second type was represented by Gram-negative, nonsporing, rod-shaped anaerobic bacteria isolated from limnic and marine mud samples with gallic acid or phloroglucinol as the sole substrate.[64] Gallic acid, pyrogallol, 1,4,6-trihydroxybenzoic acid and phloroglucinol were the only substrates utilized and were fermented to 3 mol of acetate and 1 mol of CO_2 per mole of substrate. No H_2-utilizing partner was required in these latter cultures. The latter isolate was named *Pelobacter acidigallici*. Other bacteria may convert the methoxyl group of certain aromatic compounds to acetate but do not attack the aromatic ring.[65]

2.5 PECTIN

Pectin or pectic substances are found in the primary cell wall and intercellular layers of plants.[66] Although abundant in soft tissues, only small quantities of pectins occur in woody tissues. Pectin is a group of complex plant polysaccharides chiefly containing poly D-galacturonic acid methylated in the carboxyl position. Other component sugars are D-galactose, L-arabinose, D-xylose, L-rhamnose and L-fucose.[66] Most commonly, pectin is a linear chain of α-1,4-linked-D-galacturonic acid residues with methylated carboxyl groups. The class of enzymes involved in the degradation of pectic substances is well studied and has been recently reviewed.[67] Pectinolytic rumen bacteria include *Lachnospira multiparus*,[68,69] *Bacteroides fibrisolvens*,[17] *Bacteroides ruminicola*,[70] *Bacteroides succinogenes*,[10] the rumen spirochetes[51,71] and the large rumen treponemes.[72,73] Pectinolytic *Clostridium* spp. were isolated from soil,[74] wetwood of diseased trees,[75] and lake sediment.[76] A facultative anaerobe, *Enterobacter cloacae*, obtained from Sitka spruce logs submerged in water was highly pectinolytic.[77] Three different enzymes are responsible for the hydrolysis of pectin into its component monomers: (1) pectin pectyl-hydrolase (pectinesterase) demethylates pectin to form methanol and poly D-galacturonic acid; (2) poly (α-1,4-D-galacturonide) glycanohydrolase catalyzes the hydrolytic cleavage of the α-1,4-D-galacturonan into small oligosaccharides; (3) the third enzyme, poly (α-1,4-D-galacturonide) lyase, depolymerizes the pectin chain by a trans-elimination reaction to form unsaturated galacturonic acid residues. Pectinesterase occurs in the rumen spirochetes and treponemes,[71–73] *Clostridium butyricum*,[75] unidentified *Clostridium* spp.,[74] and *Clostridium multifermentans*.[78] Polygalacturonase activity was not detected in the rumen spirochetes[71] but was present as an extracellular enzyme in rumen treponemes.[72,73] *Enterobacter cloacae* had an intracellular as well as an extracellular polygalacturonase.[77] Endopolygalacturonate lyase, perhaps the most important of the three enzymes, was detected in the extracellular culture liquid of rumen spirochetes,[71] rumen treponemes,[72,73] pigmented *Clostridium* spp.,[74] *Clostridium butyricum*,[75] and *Clostridium multifermentans*.[78] However, in *Enterobacter cloacae*, activity was intracellular.[77] Endopolygalacturonate lyase is most active between pH 8 and 9.[72,75] Efficient depolymerization of pectin requires both pectinesterase and endopolygalacturonate lyase. Methylated poly α-1,4-D-galacturonans are more difficult to depolymerize by endopolygalacturonate lyase, as are the demethylated compounds. The resultant small oligosaccharides are unsaturated. These enzymes catalyze a random hydrolysis and hence were referred to as endoenzymes in order to distinguish them from others (the exoenzymes) which attack the polymers at the terminal residue. Exopolygalacturonate lyase and exopolygalacturonase activities appear to be relatively unimportant.

2.6 STARCH

Starch is the major food reserve of higher plants, where it generally appears in the form of easily isolated granules. It is a heterogenous polysaccharide composed of two types of polymeric D-glucose. Amylose is a linear homopolysaccharide containing α-1,4 linked D-glucose residues. Amylopectin is a highly branched homopolysaccharide composed of α-1,4 linked D-glucose residues with branching via α-1,6 D-glucosidic linkages (see Greenwood[79] for details). Much is known about the enzymes and pathways of metabolism of starches in higher plants and animals, and enzymes of microbial origin were recently reviewed.[80] Starch is readily degraded in the rumen by many microorganisms which may use it as their sole energy source. *Streptococcus bovis* is present in high numbers in rumens of animals fed high grain diets, but it is not known whether it is an important starch hydrolyzer in animals on normal forage diets.[19] *Bacteroides amylophilus*[81] may be more significant under these latter conditions because of its ability to attach to starch grains. Other amylolytic rumen microbes include *Selenomonas ruminantium*, *Succinomonas amylolytica*, *Bacteroides ruminocola*,[82] the rumen spirochetes,[51] *Neocallimastix frontalis*[27] and rumen protozoa.[5] The predominant amylolytic microorganisms in digesters are rod-shaped spore-forming bacteria.[83] However, *Bacteroides* spp. are the major starch degraders in piggery-waste digesters[29,52] and *Lactobacillus* spp. are present in large numbers in anaerobic digesters treating swine waste and corn.[84] The starch-degrading enzymes may be divided into four main groups:

(1) α-Amylase (α-1,4-glucan glucanohydrolase) hydrolyzes the α-1,4 glucosidic linkage in a random manner along the starch molecule; α-1,6 linkages are not hydrolyzed. The hydrolysis products are glucose, maltose, and branched oligosaccharides. In *Bacteroides amylophilus* the α-amylase has a molecular weight of 92 000, an optimum pH of 6·3, and optimum temperature of 43 °C.[85] Ca^{2+} and Co^{2+} are strong activators of enzyme activity.

(2) α-Amylase (β-1,4 glucan maltohydrolase) cleaves maltose units from the nonreducing end of amylose or amylopectin. It is unable to pass the α-1,6 branch points in amylopectin.

(3) Amyloglucosidase releases glucose moieties from the nonreducing end of amylose or amylopectin.

(4) Debranching enzymes[82] catalyze the hydrolysis of α-1,6 linkages in amylopectin. The resulting low-molecular-weight oligosaccharides and maltose moieties are further degraded by α-glucosidase type (α-D-glucosidase glucohydrolase) enzymes. These enzymes hydrolyze the D-glucose moieties from the nonreducing end of disaccharides and oligosaccharides.

Of these four main types of enzymes, only α-amylase has been well described in anaerobic bacteria.

2.7 LIPIDS

Lipid consists mainly of glycerol with long-chain monocarboxylic acids esterified to two of its carbon atoms. The third carbon is attached to a more polar moiety — either phosphate (in phosphatides) or a sugar (in glycolipids). The phosphate may also be substituted. The predominant types of rumen lipids are mono- and digalactoglycerides or triglycerides, depending on the food source of the animal. The esterified long-chain fatty acids are mainly unsaturated.[86]

Lipid metabolism by rumen bacteria was recently reviewed,[6] but there are some differences between lipid metabolism there and in anaerobic digesters. In the rumen, *Anaerovibrio lipolytica*, a Gram-negative, curved rod,[87] actively hydrolyzes triglycerides but not phospho- or galactolipids. A *Butyrivibrio* sp. from the rumen deacylates galactolipids and phospholipids.[88] *Bacteroides fibrisolvens* hydrolyzes lecithin via a phospholipase A, yielding a free phospholipid.[89] This is degraded to glycerol and free fatty acids by a lysophospholipase. The free fatty acids released by lipolysis are not further degraded, but leave the rumen attached to food particles. They consist mainly of saturated fatty acids and some trans-mono-unsaturated acids. Rumen lipolytic bacteria all possess hydrogenation activity, but not all are capable of further metabolizing the hydrogenated fatty acid product.[6] In anaerobic digesters the hydrogenated fatty acids are metabolized further.[90] These bacteria from anaerobic digesters require low H_2 concentration for growth on these substrates,[92] although recently *Syntrophomonas wolfei* was shown to grow by fermenting crotonate.[91] The dominant lipolytic bacteria in digesters were vibrios[86] resembling the rumen isolate *Anaerovibrio lipolytica*.

2.8 PROTEIN

Protein may comprise more than 50% of the dry weight of bacterial cells; they are linear polymers composed of 20 different amino acids linked via peptide bonds. Two forms of protein are recognized: (1) simple proteins contain only amino acids; (2) conjugated proteins also contain organic and inorganic groups. Although the amino acids in proteins are arranged linearly, in addition to amino acids, proteins have varied three-dimensional conformations that are stabilized by hydrogen bonds and disulfide cross-linkages. The predominant proteolytic rumen bacteria were Gram-negative rods[6] belonging to the genus *Bacteroides.*[93–95] In contrast, the major proteolytic digestor bacteria are *Clostridium* spp.[96] The formation of proteolytic enzymes was studied in two rumen bacteria, *Bacteroides amylophilus*[97,98] and *Bacteroides ruminicola.*[94,99] *Bacteroides amylophilus* produces both cell-bound and soluble released enzyme. These appear to be different forms of the same enzyme; both are sensitive to serine-protease inhibitors,[98] although the soluble form has additional trypsin-like activity. In *Bacteroides ruminicola* the proteolytic enzymes had properties similar to the native enzyme of mixed rumen flora. In both *Bacteroides ruminicola* and rumen fluid the soluble proteases were relatively insensitive to serine-protease inhibitors.[100] Further, the major proteolytic enzymes were cysteine proteases associated with the bacterial coat or capsular material. Both exo- and endopeptidases hydrolyze proteins to oligopeptides and free amino acids. Oligopeptide metabolism is not well understood, though it is known that peptides are more readily assimilated by rumen bacteria than free amino acids.[6]

2.9 SUMMARY

The hydrolysis of polymeric substances in anaerobic environments is the rate-limiting step in the overall conversion of organic matter to methane.[1] The products of the hydrolytic reactions are fermented by both hydrolytic and nonhydrolytic bacteria. Media for the isolation of the major groups of fermentative bacteria from digesters have been developed.[97–99,25] Table 2.1 details the fermentation products of some hydrolytic bacteria. However, intermediates observed during the methanogenic degradation of organic matter by mixed microbial communities differ significantly from the pattern expected from pure culture studies. This important feature of anaerobic metabolism by mixed populations is discussed in a

TABLE 2.1
Metabolic Products Released During the Anaerobic Fermentation of Cellulose, Hemicellulose, and Pectin by Pure Cultures of Anaerobic Bacteria

Organisms	*Source*	*Substrate*	*Products**	*Ref.*
Bacteroides succinogenes	Rumen	Cellulose	F,A,S	16
Bacteroides fibrisolvens	Rumen	Cellulose	F,L,H_2,CO_2	17
		Hemicellulose	F,B,L,H_2,CO_2	15,17
		Pectin	F,A,B,L,M, H_2,CO_2	70
Bacteroides ruminicola	Rumen	Hemicellulose	F,A,P,S	15,49
		Pectin	F,A,P,B,L,S	70
Ruminococcus flavefaciens	Rumen	Cellulose	F,A,S,H_2,CO_2	104
Neocallimastix frontalis	Rumen	Cellulose	F,A,L,E,CO_2	25
Rumen spirochetes	Rumen	Pectin	F,A,M	51
			F,A,L,S,M	71
Rumen treponemes	Rumen	Pectin	F,A,S,M	72,73
Lachnospira multiparus	Rumen	Pectin	F,A,L,M,E, H_2,CO_2	69
Acetivibrio cellulolyticus	Digester	Cellulose	A,E,H_2,CO_2	30
Clostridium thermocellum	Digester	Cellulose	A,E,H_2,CO_2	105
Clostridium papyrosolvens	Estuarine sediment	Cellulose	F,A,L,E	36
Clostridium butyricum	Wetwood disease	Pectin	A,P,B,M,E,IP	75
	Lake sediment	Pectin	A,B,M,E, H_2,CO_2	76

*F = formate; A = acetate; P = propionate; B = butyrate; S = succinate; L = lactate; M = methanol; E = ethanol; IP = isopropanol.

later chapter. The importance of a full understanding of the hydrolytic reactions cannot be overemphasized: this is the primary step in anaerobic digestion and one that controls the rate of the process. Unfortunately, our knowledge of the microorganisms is limited mainly to information from the rumen system. There is a paucity of information on the hydrolytic microorganisms in anaerobic digesters, and consequently little is known about their enzymes. However, the basic mode of action of hydrolytic enzymes has been investigated in aerobic bacteria and fungi and may also apply to the anaerobic organisms. Thus, anaerobic polymeric degradation also requires the action of several hydrolytic enzymes to degrade the linear backbone of the polymer internally (endo-acting) as well as from the terminal groups.

The major organic constituents of biomass are cellulose, hemicellulose, and lignin, bound together in a lignocellulose matrix. Lignin is not only refractory to anaerobic decomposition, but also affects degradation of cellulose, hemicellulose, and other polymers. Extracellular enzymes must migrate into pore spaces (too small for bacteria) of lignocellulosic compounds to bring about hydrolysis. Such factors affecting the slow rates of lignocellulose hydrolysis are very poorly understood. Research is needed on hydrolytic bacteria and their enzymes, and the relationship between interpolymeric structure and their activities and kinetics. The effects of physical and chemical factors (temperature, pH, mixing, diffusion, enzyme inactivation, reversible and irreversible adsorption of enzymes on nonactive sites, etc.) need investigations.

A better understanding of the physiology and biochemistry of the anaerobic hydrolytic bacteria may permit better digester and reactor designs to maximize the rate of degradation of polymeric compounds. Much information on such organisms may be obtained by focusing directly on their isolation and study. The microbiology and biochemistry of pure and mixed cultures derived from prototype feed digesters should be investigated and the results related to observations on digester performance. Elucidation of the microbiology of anaerobic digestion should approach the process as an ecosystem with identification of key microbial activities and their limitations and interactions. The general approach for isolation and study of important organisms is presented below and should be done for the major groups of bacteria identified in this review, that is, hydrolytic, transitional, and methanogenic.

(1) Isolation and characterization of bacteria present in high dilutions from digesters operated on different substrates and under different conditions (loading, retention time, temperature, etc.).
(2) Study of activities (for example, substrates and products) of isolates present in high numbers and the relationship of activities to those observed in mixed culture.
(3) Determination of factors limiting conversion yields and rates of important isolates.
(4) Determination of physiological requirements, limits, and establishment of optimum conditions for growth and metabolism.
(5) Establishment of relationship of isolates with other groups of bacteria involved in the fermentation using co-culture and culture reconstruction techniques.

(6) Elucidation of important biochemical pathways and their regulation, and study of enzymes and cofactors.
(7) Use of genetic techniques for development and isolation of mutants with improved activity.

REFERENCES

1. Boone, D. R., 'Terminal Reactions in Anaerobic Digestion of Animal Waste,' *Appl. Environ. Microbiol.* **43**:57–64, 1982.
2. Henson, J. M., Bordeaux, F. M., Rivard, C. J. and Smith, P. H., 'Quantitative Influences of Butyrate or Propionate on the Thermophilic Production of Methane from Biomass,' *Appl. Environ. Microbiol.* **51**:288–292, 1986.
3. Sleat, R., Mah, R. A. and Robinson, R., 'Isolation and Characterization of an Anaerobic, Cellulolytic Bacterium, *Clostridium cellulovorans* sp. Nov.,' *Appl. Environ. Microbiol.* **48**:88–93, 1984.
4. Sleat, R., and Mah, R. A., '*Clostridium populeti* sp. Nov., a Cellulolytic Species from a Woody-Biomass Digestor,' *Int. J. Syst. Bacteriol.* **35**:160–163, 1985.
5. Hobson, P. N., and Wallace, R. J., 'Microbial Ecology and Activities in the Rumen: Part 1.' *CRC Crit. Rev. Microbiol.* **9**:162–225, 1982.
6. Hobson, P. N., and Wallace, R. J., 'Microbial Ecology and Activities in the Rumen: Part 2.' *CRC Crit. Rev. Microbiol.* **9**:253–320, 1982.
7. Wiegel, J., 'Ethanol From Cellulose,' *Experientia* **38**:151–56, 1982.
8. Zeikus, J. G., 'Chemical and Fuel Production by Anaerobic Bacteria,' *Ann. Rev. Microbiol.* **34**:423–64, 1980.
9. Halliwell, G., 'Microbial β-Gluconases,' *Prog. Ind. Microbiol.* **15**:1–62, 1979.
10. Ward, K., and Seib, P. A., 'Cellulose, Lichenin and Chitin,' pp. 413–15 in W. Pigman and D. Horton, Eds., *The Carbohydrates, Chemistry and Biochemistry*. New York and London: Academic Press, 1970.
11. Lee, Y.-H., and Fan, L. T., 'Properties and Mode of Action of Cellulase,' *Adv. in Biochem. Engr.* **17**:101–29, 1980.
12. Hungate, R. E., 'The Anaerobic Mesophilic Cellulolytic Bacteria,' *Bacteriol. Rev.* **14**:1–49, 1950.
13. Hungate, R. E., 'Studies on Cellulose Fermentation. III. The Culture and Isolation of Cellulose-Decomposing Bacteria From the Rumen of the Cattle,' *J. Bacteriol.* **53**:631–45, 1945.
14. Bryant, M. P., and Doetsch, R. N., 'A Study of Actively Cellulolytic Rod-Shaped Bacteria of the Bovine Rumen,' *J. Dairy Sci.* **37**:1176–1183, 1954.
15. Dehority, B. A., and Scott, H. W., 'Extent of Cellulose and Hemicellulose Digestion in Various Forages by Pure Cultures of Rumen Bacteria,' *J. Dairy Sci.* **50**:1136–41, 1967.

16. Stewart, C. S., Paniayua, C., Dinsdale, D., Cheng, K.-J., and Garrow, S. H., 'Selective Isolation and Characterization of *Bacteroides succinogenes* from the Rumen of a Cow,' *Appl. Environ. Microbiol.* **41**:504–10, 1981.
17. Bryant, M. P. and Small, N., 'The Anaerobic Monotrichous Butyric Acid-Producing Curved Rod-Shaped Bacteria of the Rumen,' *J. Bacteriol.* **72**:16–21, 1956.
18. Shane, B. S., Gouws, L., and Kistner, A., 'Cellulolytic Bacteria Occurring in the Rumen of Sheep Conditioned to Low-Protein Teff Hay,' *J. Gen. Microbiol.* **55**:445–51, 1969.
19. Hungate, R. E., 'Microorganisms in the Rumen of Cattle Fed a Constant Ration,' *Can. J. Microbiol.* **3**:289–311, 1957.
20. Hungate, R. E., 'Studies on Cellulose Fermentation. I. The Culture and Physiology of an Anaerobic Cellulose Digesting Bacterium,' *J. Bacteriol.* **48**:499–513, 1944.
21. Chung, K.-T., 'Inhibitory Effects of H_2 on Growth of *Clostridium cellobioparum*,' *Appl. Environ. Microbiol.* **31**:342–48, 1976.
22. Prins, R. A., van Vugt, F., Hungate, R. E. and van Vorstenbosch, C. J. A. H. V., 'A Comparison of Strains of *Eubacterium cellulosolvens* from the Rumen,' *Antonie van Leeuwenhoek* **38**:153–61, 1972.
23. Lee, B. H., and Blackburn, T. H., 'Cellulase Production by a Thermophilic *Clostridium* Species,' *Appl. Microbiol.* **30**:346–53, 1975.
24. McBee, R. H., 'The Anaerobic Thermophilic Cellulolytic Bacteria,' *Bacteriol. Rev.* **14**:51–63, 1950.
25. Bauchop, T., and Mountfort, D. O., 'Cellulose Fermentation by a Rumen Anaerobic Fungus in Both the Absence and the Presence of Rumen Methanogens,' *Appl. Environ. Microbiol.* **42**:1103–1110, 1981.
26. Mountfort, D. O., Asher, R. A. and Bauchop, T., 'Fermentation of Cellulose to Methane and Carbon Dioxide by a Rumen Anaerobic Fungus in Triculture With *Methanobrevibacter* sp. Strain RA1 and *Methanosarcina barkeri*,' *Appl. Environ. Microbiol.* **44**:128–34, 1982.
27. Orpin, C. G. and Letcher, A. J., 'Utilization of Cellulose, Starch, Xylan and Other Hemicelluloses for Growth by the Rumen Phycomycète *Neocallimastix frontalis*,' *Curr. Microbiol.* **3**:121–24, 1979.
28. Maki, L. R., 'Experiments on the Microbiology of Cellulose Decomposition in a Municipal Sewage Plant,' *Antonie van Leeuwenhoek* **20**:185–200, 1954.
29. Iannotti, E. L., Fischer, J. R., and Sievers, D. M., 'Characterization of Bacteria From a Swine Manure Digester,' *Appl. Environ. Microbiol.* **43**:136–43, 1982.
30. Patel, G. B., Khan, A. W., Agnew, B. J. and Colcin, J. R., 'Isolation and Characterization of an Anaerobic, Cellulolytic Microorganism, *Acetivibrio cellulolyticus* gen. nov. sp. nov.,' *Int. J. System. Bacteriol.* **30**:179–85, 1980.
31. Betaine, H. G., Linehan, B. A., Bryant, M. P. and Holdeman, L. V., 'Isolation of a Cellulolytic Bacteroides sp. From Human Feces,' *Appl. Environ. Microbiol.* **33**:1009–10, 1977.
32. Montgomery, L. and Macy, J. M., 'Characterization of Rat Cecum Cellulolytic Bacteria,' *Appl. Environ. Microbiol.* **44**:1435–43, 1982.
33. Davies, M. E., 'Cellulolytic Bacteria Isolated From the Large Intestine of the Horse,' *J. Appl. Bacteriol.* **27**:373–78, 1964.

34. Dehority, B. A., 'Cellulolytic Cocci Isolated From the Cecum of Guinea Pigs (*Cavia porcellus*),' *Appl. Environ. Microbiol.* **33**:1278–83, 1977.
35. Skinner, F. A., 'The Isolation of Anaerobic Cellulose-Decomposing Bacteria From Soil,' *J. Gen. Microbiol.* **22**:53–54, 1960.
36. Madden, R. H., Bryder, M. J. and Poole, N. J., 'Isolation and Characterization of an Anaerobic, Cellulolytic Bacterium, *Clostridium papyrosolvens* sp. nov.,' *Int. J. System. Bacteriol.* **32**:87–91.
37. Groleau, D and Forsberg, C. W., 'Cellulolytic Activity of the Rumen Bacterium *Bacteroides succinogenes*,' *Can. J. Microbiol.* **27**:517–530, 1981.
38. Latham, M. J., Booker, B. E., Pettipher, P. L. and Harris, P. J., 'Adhesion of *Bacteroides succinogenes* in Pure Culture and in the Presence of *Ruminococcus flavefaciens* to Cell Walls in Leaves of Perennial Ryegrass (*Lolium perennne*),' *Appl. Environ. Microbiol.* **35**:156–165, 1978.
39. Johnson, E. A., Sakajoh, M., Halliwell, Madia A. and Demain, A. L., 'Saccharification of Complex Cellulosic Substrates by the Cellulase System From *Clostridium thermocellum*,' *Appl. Environ. Microbiol.* **43**:1125–32, 1981.
40. Ng, T. K., Weimer, P. J. and Zeikus, J. G., 'Cellulolytic and Physiological Properties of *Clostridium thermocellum*,' *Arch. Microbiol.* **114**:1–7, 1977.
41. Ng, T. K. and Zeikus, J. G., 'Comparison of Extracellular Cellulase Activities of *Clostridium thermocellum* LQRI and *Trichoderma reesei* QM9414,' *Appl. Environ. Microbiol.* **42**:231–40, 1981.
42. Fusee, M. C., and Heatherwood, J. M., 'Regulation of Cellulase From *Ruminococcus*,' *Can. J. Microbiol.* **18**:347–53, 1972.
43. Pettipher, G. L. and Latham, M. J., 'Characterization of Enzymes Produced by *Ruminococcus flavefaciens* Which Degrade Plant Cell Walls,' *J. Gen. Microbiol.* **110**:21–27, 1979.
44. Smith, W. R., Yu, I. and Hungate, R. E., 'Factors Affecting Cellulolysis by *Ruminococcus albus*,' *J. Bacteriol.* **114**:729–37, 1973.
45. Saddler, J. N. and Khan, A. W., 'Cellulase Production by *Acetivibrio cellulolyticus*,' *Can. J. Microbiol.* **26**:760–65, 1980.
46. Gong, C.-S., Chu, L. F., Flickinger, M. C. and Tsao, G. T., 'Conversion of Hemicellulose Carbohydrates,' *Adv. Biochem. Eng.* **20**:93–118, 1981.
47. Whistler, R. L. and Richards, E. L., 'Hemicelluloses,' pp. 447–69 in W. Pigman and D. Horeton, Eds., *The Carbohydrates, Chemistry and Biochemistry*. New York and London: Academic Press, 1970.
48. Dekker, R. F. H. and Richards, G. R., 'Hemicellulases: Their Occurrence, Purification, Properties and Mode of Action,' *Adv. Carbohydrate Chem. Biochem.* **32**:277–352, 1976.
49. Dehority, B. A., 'Characterization of Several Bovine Rumen Bacteria Isolated With a Xylan Medium,' *J. Bacteriol.* **91**:1724–29, 1966.
50. Dehority, B. A., 'Degradation and Utilization of Isolated Hemicellulose by Pure Cultures of Cellulolytic Rumen Bacteria,' *J. Bacteriol.* **89**:1515–20, 1965.
51. Paster, P. J. and Canale-Parola, E., 'Physiological Diversity of Rumen Spirochetes,' *Appl. Environ. Microbiol.* **43**:686–93, 1982.
52. Hobson, P. N. and Shaw, B. G., 'The Bacterial Population of Piggery-Waste Anaerobic Digesters,' *Water Res.* **8**:507–516, 1974.

53. Salyers, A. A., West, S. E. H., Vercellotti, J. R. and Wilkins, T. D., 'Fermentation of Mucins and Plant Polysaccharides by Strains of *Bacteroides* from the Human Colon,' *Appl. Environ. Microbiol.* **33**:319–322, 1977.
54. Salyers, A. A., West, S. E. H., Vercellotti, J. R. and Wilkins, T. D., 'Fermentation of Mucins and Plant Polysaccharides by Strains of Anaerobic Bacteria From the Human Colon,' *Appl. Environ. Microbiol.* **33**:529–33, 1977.
55. Salyers, A. A., Ghardini, F. and O'Brien, M., 'Utilization of Xylan by Two Species of Human Colonic Bacteroides,' *Appl. Environ. Microbiol.* **41**:1065–68, 1981.
56. Reddy, C. A., 'Introduction to Microbial Degradation of Lignins,' *Dev. Ind. Microbiol.* **19**:23–26, 1978.
57. Hackett, W. F., Connors, W. J., Kirk, T. K. and Zeikus, J. G., 'Microbial Decomposition of Synthetic ^{14}C-Labelled Lignins in Nature: Lignin Biodegradation in a Variety of Natural Materials,' *Appl. Environ. Microbiol.* **33**:43–51, 1977.
58. Colberg, P. J. and Young, L. Y., 'Biodegradation of Lignin-Derived Molecules Under Anaerobic Conditions,' *Can. J. Microbiol.* **28**:886–89, 1982.
59. Healy, J. B. and Young, L. Y., 'Catechol and Phenol Degradation by a Methanogenic Population of Bacteria,' *Appl. Environ. Microbiol.* **35**:216–18, 1978.
60. Healy, J. B. and Young, L. Y., 'Anaerobic Degradation of Eleven Aromatic Compounds to Methane,' *Appl. Environ. Microbiol.* **38**:84–89, 1979.
61. Healy, J. B., Young, L. Y. and Reinhard, M., 'Methanogenic Decomposition of Ferulic acid, A Model Lignin Derivative,' *Appl. Environ. Microbiol.* **39**:436–44, 1980.
62. Zeikus, J. G., Wellstein, A. L. and Kirk, T. K., 'Molecular Basis for the Biodegradative Recalcitrance of Lignin in Anaerobic Environments,' *FEMS Microbiol. Lett.* **15**:193–97, 1982.
63. Mountfort, D. O. and Bryant, M. P., 'Isolation and Characterization of an Anaerobic Syntrophic Benzoate Degrading Bacterium From Sewage Sludge,' *Arch. Microbiol.* **133**:249–56, 1982.
64. Schink, B. and Pfennig, N., 'Fermentation of Trihydroxybenzenes by *Pelobacter acidigallici* Gen. Nov. Sp. Nov., a New Strictly Anaerobic, Non-Sporeforming Bacterium,' *Arch. Microbiol.* **133**:195–201, 1982.
65. Bache, R. and Pfennig, N., 'Selective Isolation of *Acetobacterium woodii* on Methoxylated Aromatic Acids and Determination of Growth Yields,' *Arch. Microbiol.* **130**:255–61, 1981.
66. Aspinall, G. O., 'Pectins, Plant Gums, and Other Plant Polysaccharides,' pp. 515–36 in W. Pigman and D. Horton, Eds., *The Carbohydrates, Chemistry and Biochemistry*. New York and London: Academic Press, 1970.
67. Rexova-Benkova, L. and Markovic, O., 'Pectic Enzymes,' *Adv. Carbohydrate Chem. Biochem.* **33**:323–85, 1976.
68. Bryant, M. P. and Small, N., 'Characteristics of Two New Genera of Anaerobic Curved Rods Isolated From the Rumen of Cattle,' *J. Bacteriol.* **72**:22–26, 1956.

69. Rode, L. M., Sharak-Genthner, B. R. and Bryant, M. P., 'Syntrophic Association by Cocultures of Methanol- and CO_2–H_2-Utilizing Species *Eubacterium limosum* and Pectin-Fermenting *Lachnospira multiparus* During Growth in Pectin Medium,' *Appl. Environ. Microbiol.* **42**:20–22, 1981.
70. Dehority, B. A., 'Pectin Fermenting Bacteria Isolated From the Bovine Rumen,' *J. Bacteriol.* **99**:189–96, 1969.
71. Ziolecki, A. and Wojckiechowicz, M., 'Small Pectinolytic Spirochetes From the Rumen,' *Appl. Environ. Microbiol.* **39**:919–22, 1980.
72. Wojckiechowicz, M. and Ziolecki, A., 'Pectinolytic Enzymes of the Large Rumen Treponemes,' *Appl. Environ. Microbiol.* **37**:136–42, 1979.
73. Ziolecki, A., 'Isolation and Characterization of Large Treponemes From the Bovine Rumen,' *Appl. Environ. Microbiol.* **37**:131–35, 1979.
74. Lund, B. M. and Brocklehurst, T. F., 'Pectic Enzymes of Pigmented Strains of *Clostridium*,' *J. Gen. Microbiol.* **104**:59–66, 1978.
75. Schink, B., Ward, J. C. and Zeikus, J. G., 'Microbiology of Wetwood: Importance of Pectin Degradation and *Clostridium* Species in Living Trees,' *Appl. Environ. Microbiol.* **42**:526–32, 1981.
76. Schink, B. and Zeikus, J. G., 'Microbial Ecology of Pectin Decomposition in Anoxic Sediments,' *J. Gen. Microbiol.* **128**:393–404, 1982.
77. Macken, J. and Pickaver, A. H., 'Synthesis of Polygalacturonate Trans-Eliminase and Polygalacturonase by a Strain of *Enterobacter cloacae* Isolated From Ponded Sitka Spruce,' *J. Appl. Bacteriol.* **46**:75–86, 1979.
78. Sheiman, M. J., MacMillan, J. D., Miller, L. and Chase, T., 'Coordinated Action of Pectinesterase and Polygalacturonate Lyase Complex of *Clostridium multifermentans*,' *Eur. J. Biochem.* **64**:565–72, 1976.
79. Greenwood, C. T., 'Starch and Glycogen,' pp. 471–513 in W. Pigman and D. Horton, Eds., *The Carbohydrates, Chemistry and Biochemistry*. New York and London: Academic Press, 1970.
80. Fogarty, W. M. and Kelly, C. T., 'Starch Degrading Enzymes of Microbial Origin,' *Prog. Ind. Microbiol.* **15**:87–150.
81. Hamlin, L. J. and Hungate, R. E., 'Culture and Physiology of a Starch Degrading Bacterium (*Bacteroides amylophilus N.* sp.) from the Bovine Rumen,' *J. Bacteriol.* **72**:548–54, 1956.
82. Hungate, R. E., 'Microbial Ecology of the Rumen,' *Bacteriol. Rev.* **24**:353–64, 1960.
83. Torien, D. F., 'Enrichment Culture Studies on Aerobic and Anaerobic Bacteria Found in Anaerobic Digesters,' *Water Res.* **1**:147–55, 1967.
84. Nakumura, L. K. and Crowell, C. D., '*Lactobacillus amylophilus*, A New Starch-Hydrolyzing Species From Swine-Waste Corn Fermentation,' *Dev. Ind. Microbiol.* **20**:531–40, 1979.
85. McWethy, S. J. and Hartman, P. A., 'Purification and Some Properties of an Extracellular Alpha-Amylase From *Bacteroides amylophilus*,' *J. Bacteriol.* **129**:1537–44, 1977.
86. Hobson, P. N., Bousfield, S. and Summers, R., 'Anaerobic Digestion of Organic Matter,' *CRC Crit. Rev. Environ. Control* **4**:131–91, 1974.
87. Hobson, P. N. and Mann, S. O., 'The Isolation of Glycerol-Fermenting and

Lipolytic Bacteria From the Rumen of the Sheep,' *J. Gen. Microbiol.* **25**:227–40, 1961.

88. Hazelwood, G. P. and Dawson, R. M. C., 'Characteristics of a Lipolytic and Fatty Acid Requiring *Butyrivibrio* sp. Isolated From the Bovine Rumen,' *J. Gen. Microbiol.* **112**:15–27, 1979.
89. Hazelwood, G. P. and Dawson, R. M. C., 'Isolation and Properties of a Phospholipid-Hydrolyzing Bacterium From Ovine Rumen Fluid,' *J. Gen. Microbiol.* **89**:163–74, 1975.
90. Chynoweth, D. P. and Mah, R. A., 'Volatile Acid Formation in Sludge Digestion,' *Adv. Chem. Ser.* **105**:41–54, Amer. Chem. Soc., 1971.
91. Beaty, P. S. and McInerney, M. J., 'Isolation of *Syntrophomonas wolfei* on Crotonate,' *Abst. Ann. Meeting Am. Soc. Microbiol.*, 1132, 1986.
92. McInerney, M. J., Bryant, M. P. and Pfennig, N., 'Anaerobic Bacterium that Degrades Fatty Acids in Syntrophic Association With Methanogens,' *Arch. Microbiol.* **122**:129–135, 1979.
93. Blackburn, T. H. and Hobson, P. N., 'Further Studies on the Isolation of Proteolytic Bacteria From the Sheep Rumen,' *J. Gen. Microbiol.* **29**:69–81, 1962.
94. Hazelwood, G. P., Jones, G. A and Mangan, J. L., 'Hydrolysis of Leaf Fraction 1 Protein by the Proteolytic Rumen Bacterium *Bacteroides ruminicola* R8/4,' *J. Gen. Microbiol.* **123**:223–32, 1981.
95. Hazelwood, G. P. and Nugent, J. H. A., 'Leaf Fraction 1 Protein as a Nitrogen Source for the Growth of a Proteolytic Rumen Bacterium,' *J. Gen. Microbiol.* **106**:369–71, 1978.
96. Siebert, M. L. and Torien, D. F., 'The Proteolytic Bacteria Present in the Anaerobic Digestion of Raw Sewage Sludge,' *Water Res.* **3**:241–50, 1969.
97. Blackburn, T. H., 'Protease Production by *Bacteroides amylophilus* Strain H18,' *J. Gen. Microbiol.* **53**:27–36, 1968.
98. Blackburn, T. H., 'The Protease Liberated From *Bacteroides amylophilus* Strain H18 by Mechanical Disintegration,' *J. Gen. Microbiol.* **53**:37–51, 1968.
99. Hazelwood, G. P. and Edwards, R., 'Proteolytic Activities of a Rumen Bacterium, *Bacteroides ruminicola* R8/4,' *J. Gen. Microbiol.* **125**:11–15, 1981.
100. Kopecny, J. and Wallace, R. J., 'Cellular Location and Some Properties of Proteolytic Enzymes of Rumen Bacteria,' *Appl. Environ. Microbiol.* **43**:1026–33, 1982.
101. Mah, R. A. and Sussman, C., 'Microbiology of Anaerobic Sludge Fermentation. I. Enumeration of the Nonmethanogenic Anaerobic Bacteria,' *Appl. Microbiol.* **16**:358–61, 1968.
102. Kirsch, E. J., 'Studies on the Enumeration and Isolation of Obligate Anaerobic Bacteria From Digesting Sewage Sludge,' *Dev. Ind. Microbiol.* **10**:170–76, 1969.
103. Torien, D. F., 'Population Description of the Non-Methanogenic Phase of Anaerobic Digestion. I. Isolation, Characterization and Identification of Numerically Important Bacteria,' *Water Res.* **4**:136–43, 1970.
104. Latham, M. J. and Wolin, M. J., 'Fermentation of Cellulose by *Rumino-*

coccus flavefaciens in the Presence and Absence of *Methanobacterium ruminantium*,' *Appl. Environ. Microbiol.* **34**:297–301, 1977.

105. Weimer, P. J. and Zeikus, J. G., 'Fermentation of Cellulose and Cellobiose by *Clostridium thermocellum* in the Absence and Presence of *Methanobacterium thermoautotrophicum*,' *Appl. Environ. Microbiol.* **33**:289–297, 1977.

3
Transitional Bacteria

DAVE BOONE and ROBERT MAH

School of Public Health, University of California, Los Angeles, California, USA

3.1 INTRODUCTION

Transitional bacteria convert the soluble organic matter produced by hydrolytic bacteria into methanogenic substrates. Soluble organic matter enters the anaerobic degradation process at a number of different points (Fig. 3.1). For instance, acetate in the influent may be metabolized directly by methanogens independent of catabolic interactions with other bacteria. Fatty acids, the major hydrolysis products of fats, are catabolized primarily by the obligate proton-reducing acetogens and converted to methanogenic substrates.[1] Peptides, the hydrolysis products of protein, are taken up by bacteria and hydrolyzed to amino acids that may be used as carbon and energy sources via fermentative reactions. Oligosaccharides, the major hydrolysis products of carbohydrate polymers, are taken up by bacteria and catabolized by various fermentative pathways. Traditionally, anaerobic digestion has been described as a two-phase process with the products of the first or acidogenic phase serving as intermediate substrates for the second or methanogenic phase. Some of the intermediates in question (for example, acetate and hydrogen) are used directly by methanogenic bacteria, but others (such as volatile fatty acids of chain length greater than acetate) must first be catabolized to compounds which can be used by methanogens. A recently discovered group of unique bacteria, the obligate proton-reducing acetogens, is responsible for this catabolism and has led to the development of a new model (Fig. 3.1). The two-phase model of anaerobic digestion highlights the important distinction between acidogenic bacteria and bacteria involved in subsequent degradations, but its disadvantage is that it obscures the important metabolic interactions between acidogenic (fermentative) bacteria and

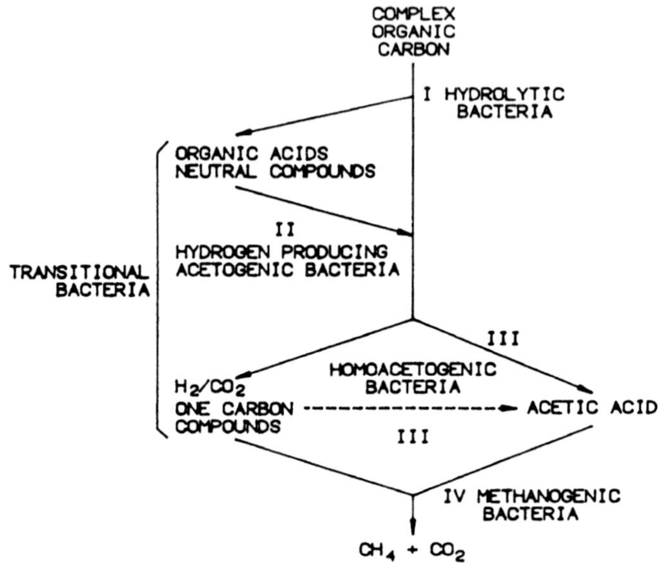

FIG. 3.1. Categories of metabolically distinct bacteria in the methane fermentation.

the H_2-utilizing methanogens, and the interactions between the obligate proton-reducing acetogens and methanogens. Both the obligate proton-reducing acetogenic and fermentative chemoheterotrophic bacteria produce the methanogenic substrates, acetate and H_2–CO_2. The important distinction between these two types of heterotrophs is the ability of fermentative bacteria to grow whether or not the H_2 they produce is oxidized and removed. In the absence of hydrogen removal, electrons normally channelled to hydrogen production are disposed of by other pathways.[2] The fermentative chemoheterotrophs often metabolize a wide range of carbohydrate substrates; in anaerobic digestors, they produce mainly volatile fatty acids, H_2 and CO_2. The obligate proton-reducing acetogens, on the other hand, form acetate and grow only by producing H_2, which they make only when the H_2 concentration is low. Therefore, these latter organisms require a very tight metabolic interaction with H_2-utilizing bacteria. These organisms are important in anaerobic digesters, because acetate, propionate, and butyrate are intermediates in the conversion of the vast majority of organic substrates to CH_4 and CO_2. These are generally measured by the isotopic turnover technique;[3,4]

propionate turnover should be measured with [2-^{14}C]- or [3-^{14}C]-propionate because the carboxyl group may exchange with CO_2.[5]

A third group of organisms (including homoacetogenic bacteria) which catalyze 'back reactions' (such as acetate production from H_2 and CO_2) will also be discussed in this chapter. The interaction of H_2-producing and H_2-oxidizing bacteria is central to the control of the methane fermentation and is referred to as interspecies H_2 transfer. The first methanogenic interspecies H_2-transfer couplet described was the ethanol-fermenting culture formerly known as *Methanobacillus omelianskii*. In an often-cited classical study, Bryant *et al.*[6] found that '*Methanobacillus omelianskii*', presumed to be axenic, actually consisted of a symbiotic co-culture of two bacterial species. When the co-culture was transferred from ethanol to H_2–CO_2 medium, it irreversibly lost the ability to utilize ethanol as a methanogenic substrate. This result, together with the observation of two cell morphotypes in the ethanol cultures, suggested the 'isolate' was a co-culture of a nonmethanogenic bacterium and a hydrogen-oxidizing methanogen. Bryant separated the organisms into two axenic cultures. The nonmethanogen oxidized ethanol to acetate and H_2; the methanogen oxidized H_2 to form CH_4 via CO_2 reduction. They were mutually dependent when grown on ethanol since the nonmethanogen was inhibited by the H_2 that was removed by the methanogen. The latter depended on the H_2 formed by the ethanol-oxidizing partner.

3.2 FERMENTATIVE BACTERIA

Fermentative bacteria in anaerobic digesters convert soluble organic material mainly to acetate, propionate, butyrate, H_2, and CO_2, presumably via known pathways of metabolism. The electrons generated during the fermentation reactions may be transferred to pyridine nucleotides and ultimately disposed of via proton reduction to form H_2. Certain fermentative reactions proceed only a low H_2 concentrations and depend on H_2 removal by H_2-oxidizing bacteria.[7] The resulting low H_2 concentration accompanies formation of acetate as the major soluble product. Furthermore, the continuous removal of H_2 stimulates its own formation, and H_2 together with CO_2 then become important products. When H_2 removal by methanogens is less efficient, H_2 blocks electron disposal by proton reduction, and fermentative bacteria must channel electrons to other disposal sites.[2] This results in increased production of reduced fermentation products such as propionate and butyrate, and is

TABLE 3.1
Interspecies H_2-Transfer Couplets

	Substrate*	Alone*	With H_2 utilizer†	Ref.
S Organism	EtOH, CO_2,	HAc, H_2	HAc, CH_4	6
	PyA	HAc, EtOH, CO_2, H_2(tr)	HAc, CH_4, EtOH(tr)	10
Desulfovibrio	EtOH,	HAc, H_2S‡	HAc, CH_4	2
	LA	HSuc, H_2, CO_2	HAc, CH_4, CO_2	2
Ruminococcus flavefaciens	Cellulose	HAc, HCOOH, HSuc, H_2, CO_2	CH_4, HAc, HSuc	11
Selenomonas ruminantium	Glucose	HAc, HPr, CO_2, LA, H_2(tr)	CH_4, HAc, HPr, CO_2, LA	12
Clostridium cellobioparum	Glucose	HAc, EtOH, HCOOH, H_2, CO_2 (LA, HBu)	CH_4, EtOH, HAc	13
Clostridium thermocellum	Cellulose, LA, H_2, CO_2	HAc, EtOH, HBu, LA, CO_2	CH_4, EtOH, HAc	14
Ruminococcus albus	Glucose, CO_2	HAc, EtOH, H_2	HAc, HSuc,§ CO_2	7

*Abbreviations: EtOH = ethanol; PyA = pyruvic acid; HAc = acetic acid; HSuc = succinic acid; HPr = propionic acid; LA = lactic acid; HBu = butyric acid.
†Partner is H_2-utilizing methanogen, except *Ruminococcus albus* has *Vibrio succinogenes* as partners.
‡SO_4^{2-} added to medium.
§Fumarate added to medium.

called H_2 control point number 1.[7] When fermentative bacteria are isolated from digesters and grown in axenic culture, there is no mechanism for interspecies H_2 removal. Such cultures may form fermentation products such as lactate and ethanol which are not usually produced in anaerobic digesters. This observation by Hungate on the rumen system[8] was subsequently first documented by the co-culture studies of Scheifinger *et al.*[9] Axenic cultures of fermentative bacteria exhibit shifts in fermentation products when H_2-using methanogens are added (Table 3.1). In co-culture, the formation of reduced end-products (propionate, butyrate, lactate, ethanol) decreases, and H_2 becomes the major electron disposal product. H_2, however, is not detected among the products formed because it is oxidized by methanogens to reduce CO_2 to CH_4. Acetate production increases because proton reduction (that is, H_2 formation) becomes the main route of electron disposal and reduced organic end-products are not significant. Axenic cultures of *Acetobacterium woodii* produce 3 mol of acetate per mole of hexose by fermentation pathways. However, in co-culture with H_2-oxidizing methanogens, H_2 removal (due to CH_4 formation) actually results in less acetate production.

3.3 OBLIGATE PROTON-REDUCING ACETOGENS

Some of the products of fermentative bacteria, namely acetate and H_2, can be metabolized directly by methanogenic bacteria, but others, such as propionate and butyrate, cannot. Although about one-half of the biodegradable chemical oxygen demand is fermented to propionate or butyrate,[15–17] these products must be further degraded to acetate and H_2 by the concerted action of obligate proton-reducing acetogens growing with H_2-oxidizing methanogens. An examination of the standard free-energy changes of fatty acid oxidation (Table 3.2) indicates that they are endergonic at standard concentrations of substrates and products, and

TABLE 3.2
Free-Energy Changes for Several Important Reactions of Anaerobic Digestion

	kJ/reaction *G*	*G**	*Allowable H_2 partial pressure, atm*†
Methanogenesis:			
$1/4\,HCO_3^- + H_2 + 1/4\,H^+ \rightarrow 1/4\,CH_4 + 3/4\,H_2O$	−33·9	−4·51	$>1{\cdot}6 \times 10^{-6}$
Propionate oxidation:			
$1/3\,CH_3CH_2COO^- + H_2O \rightarrow 1/3\,CH_3COO^- + 1/3\,HCO_3^- + 1/3\,H^+ + H_2$	+25·4	−4·29	$<5{\cdot}8 \times 10^{-5}$
Butyrate oxidation:			
$1/2\,CH_3CH_2CH_2COO^- + H_2O \rightarrow CH_3COO^- + H_2 + 1/2\,H^+$	+24·0	−9·22	$<4{\cdot}4 \times 10^{-4}$

*Free energy changes calculated at concentrations measured or estimated in an actual anaerobic digester (acetate, 1·18 mM; propionate, 0·66 mM; butyrate, 0·10 mM; HCO_3^-, 78·1 mM).
†H_2 partial pressures above or below which the reaction has a negative free-energy change (concentrations as in above footnote, except H_2).

cannot occur under these conditions. However, since these reactions do occur in digesters as bacterial energy-yielding reactions, the substrate and product concentrations must differ from those described for standard free-energy changes in order to be feasible. Table 3.2 shows that an H_2 concentration below $5{\cdot}8 \times 10^{-5}$ atm would allow propionate oxidation to take place. At very low H_2 concentrations, however, methanogenesis from H_2 and CO_2 becomes unfavorable. For that reaction to be exergonic the H_2 partial pressure must be above $1{\cdot}6 \times 10^{-6}$ atm. Thus, H_2 partial pressures in the digester must fall somewhere between $1{\cdot}6 \times 10^{-6}$ and $5{\cdot}8 \times 10^{-5}$ atm. The regulation of H_2 on the catabolic reactions of the obligate proton-reducing acetogens is termed the 'second site of hydrogen regulation'[7] since H_2 concentration controls these reactions. McInerney *et al.*[18] unequivocally demonstrated that certain oxidation–reduction reactions that are thermodynamically unfavorable under standard conditions can occur if the H_2 concentrations are kept very low using two-membered co-cultures of bacteria. They did this by growing H_2-utilizing methanogens with *Syntrophomonas wolfei*, which oxidizes butyrate to acetate and H_2. In the presence of H_2, or when H_2 removal is inhibited, growth and butyrate oxidation by *Syntrophomonas wolfei* stop, as predicted from the thermodynamics of the reaction (Fig. 3.2). *Syntrophomonas wolfei* uses normal monocarboxylic acids (C_4 to C_8) as substrates (isoheptanoate may also be used).[19] The normal acids are β-oxidized to acetate and H_2; the odd-chained acids are β-oxidized to acetate, propionate, and H_2. None of these reactions occurs without keeping the H_2 concentration low by co-culturing with an H_2-oxidizing species. An extensive search for other substrates or other electron acceptors recently indicated that crotonate may be used by axenic cultures of *Syntrophomonas wolfei* without H_2 removal.[20] Fatty acids with more than eight carbon atoms, although major substrates in domestic sewage sludge,[21] are not degraded by *Syntrophobacter wolfei*. However, the isolation in monoxenic culture of an analogous long-chain fatty acid oxidizer has been accomplished (M. P. Bryant, personal communication). The second obligate proton-reducing acetogen isolated in monoxenic culture is *Syntrophobacter wolinii*.[22] This species oxidizes propionate to acetate, CO_2, and H_2 as its sole energy-yielding reaction. It requires very efficient H_2 removal to make the reaction exergonic. Unlike *Syntrophomonas wolfei*, no alternate growth compounds have been found, and *Syntrophobacter wolinii* is an obligate syntroph. It grows very slowly, with a minimum doubling time of 3·6 days (with sulfate reduction to sulfide as the H_2 sink) or 6·7 days (with CO_2 reduction to methane as the electron

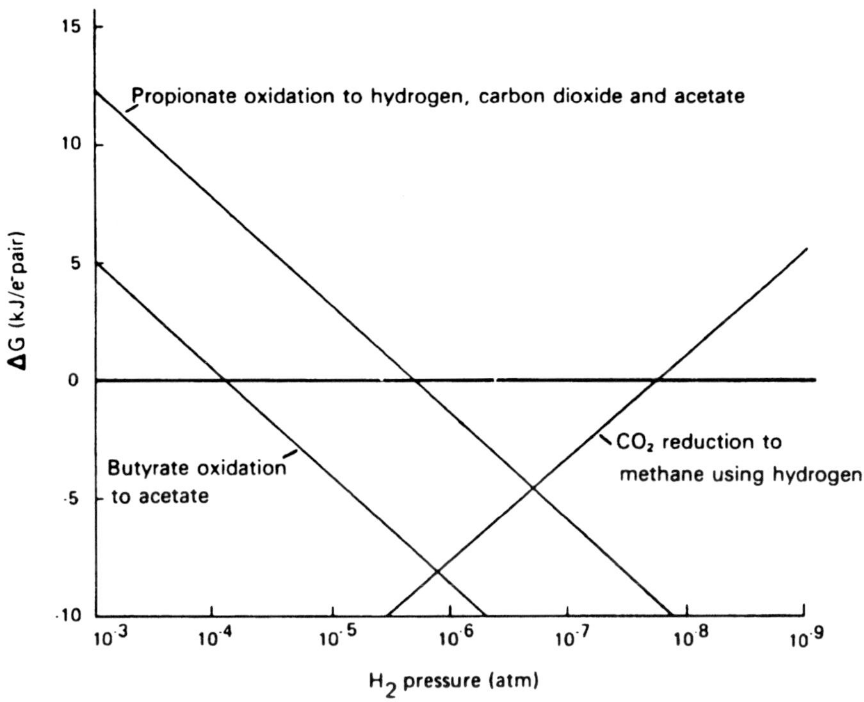

FIG. 3.2. Free-energy change for hydrogen production from propionate and butyrate, and for methanogenesis from hydrogen and bicarbonate (free-energy was calculated for reactions of Table 3.2 at various hydrogen concentrations and with other concentrations as measured, or estimated, in actual digester: acetate, 1·18 mM; propionate, 0·66 mM; butyrate, 0·40 mM (estimated); methane, 51·4 kPa; and bicarbonate, 78·1 mM).

sink). These values correspond to dilution rates above which propionate and acetate begin to accumulate in anaerobic digesters, presumably due to washout of the slow-growing propionate-degrading bacteria.[23] Recently, obligate proton-reducing acetogens were isolated in monoxenic co-culture from thermophilic systems. Henson and Smith[24] isolated a thermophilic butyrate-oxidizing bacterium in co-culture with *Methanobacterium thermoautotrophicum*. This organism oxidizes butyrate to acetate and H_2, with H_2 continuously removed by the methanogen. Zinder and Koch[25] reported the isolation in monoxenic culture of a thermophilic organism that oxidizes acetate to H_2 and CO_2 in co-culture with a thermophilic, H_2-oxidizing methanogen. This acetate-oxidizing thermophile is the only known obligate proton-reducer that does not

TABLE 3.3
Methanogenic Interspecies H_2-Transfer

Chemoheterotrophic H_2 producer	*Substrate*	*Monoculture products*	*Mixed H_2-using methanogen*	*Culture products*††	*Refs.*
Thermoanaerobium brockii	Ethanol	Acetate (tr), H_2 (tr)	*Methanobacterium thermoautotrophicum*	Acetate (I), CH_4	26
S Organism	Ethanol	Acetate (tr), H_2 (tr)	*M. bryantii*	Acetate (I), CH_4	10
	Other alcohols*	None	*Methanobrevibacter smithii*	Corresponding acid,† CH_4	10
	Pyruvate‡§	Ethanol, acetate, H_2	*M. smithii*	CH_4, acetate (I)	10
Desulfovibrio desulfuricans	Ethanol	Acetate (tr), H_2 (tr)	*M. bryantii*	Acetate (I), CH_4	27
D. vulgaris	Lactate	Trace prods.	*M. bryantii*	Acetate (I), CH_4	27
	Lactate	Trace prods.	*Methanosarcina barkerii*	CH_4, CO_2	27
Selenomonas ruminantium	Lactate§	Acetate, propionate	*M. smithii*	Acetate (I), propionate (L), CH_4	12
Anaerovibrio lipolytica	Lactate§	Acetate, propionate, H_2 (tr)	*M. ruminantium*	Acetate (I), propionate, CH_4	28

Syntrophomonas wolfei	Butyrate	None	*Methanospirillum hungatei***	Acetate, CH_4	18
	Valerate	None	*M. hungatei*	Acetate, propionate, CH_4	
Syntrophobacter wolinii	Propionate	None	*Desulfovibrio* sp. + sulfate	Acetate, H_2S	22
	Propionate	None	*Desulfovibrio* sp. + *M. hungatei*	Acetate, CH_4	22

*Other alcohols: *n*-propanol, *n*-butanol, isobutanol, *n*-pentanol. Acetaldehyde was also a substrate.
†Presumably the corresponding fatty acid was formed.
‡Oxalacetate was also a fermentative substrate for S organism.
§These substrates are metabolized fermentatively by axenic cultures of the chemoheterotroph shown; proton-reduction is not obligate.
***Methanobacterium bryantii* and *Methanobrevibacter arboriphilus* were also tested but did not support growth comparable with *M. hungatei*.
††(I) indicates increase in product; (L) indicates decrease in product.

produce acetate. Although the oxidation of acetate to CO_2 may occur to a limited extent in digestors, it is not a significant reaction there.[4] Most of the acetate in digesters is cleaved directly by methanogens to CH_4 and CO_2. Table 3.3 summarizes the obligate proton-reducing acetogens currently in monoxenic culture. Included is a species that degrades aromatic compounds, as discussed in Chapter 2.

3.4 HYDROGEN-USING 'BACK REACTIONS'

Labeling studies indicate that a small amount of acetate is synthesized from H_2 and CO_2 in digesters and that a small amount is also converted to propionate or butyrate.[21,4] Other studies indicate that mixed cultures produce volatile fatty acids from H_2 and CO_2[29–31] or from methanol.[32,33] Several organisms that carry out these reactions have been isolated (Table 3.4). *Eubacterium limosum* is capable of producing volatile acids of chain length greater than acetate, starting with C_1 compounds.[34] The culture, which is called '*Butyribacterium methylotrophicum*',[35] is probably a strain of *Eubacterium limosum* but was named before *Eubacterium limosum* was known to metabolize C_1 compounds. *Eubacterium limosum* may be numerous in sludge digesters or in the rumen. Other organisms that produce volatile fatty acids solely from C_1 compounds are *Peptostreptococcus productus* and *Acetobacterium woodii* (Table 3.4). *Peptostreptococcus productus* can grow on CO or H_2–CO_2, producing acetate and succinate.[36] *Desulfobulbus propionicus* produces propionate

TABLE 3.4
Bacteria Catalyzing Back Reactions

Organism	*Products*	*Refs.*
Eubacterium limosum	Acetate, butyrate	35
Peptostreptococcus productus	Acetate, succinate	37
Desulfobulbus propionicus	Propionate	40
Acetobacterium woodii	Acetate	41
Clostridium thermoaceticum	Acetate	42, 43
Acetobacterium wieringae	Acetate	44
Clostridium aceticum	Acetate	45, 46
Clostridium formicoaceticum	Acetate	47
Acetobacterium kivui	Acetate	44

from acetate and H_2 or from ethanol.[37] Other organisms shown in Table 3.4 produce acetate from CO_2 only when grown on certain carbohydrates. The production of volatile fatty acids from H_2 and CO_2 or H_2, CO_2, and acetate are the reverse of reactions carried out by the obligate proton-reducing acetogens. Therefore, both kinds of reactions cannot be exergonic under the same conditions. However, since H_2 has a rapid turnover, it is likely that H_2 concentrations may differ significantly at the microenvironmental level.[38,39] Under these conditions, the H_2-producing acetogenic reactions and the 'back reactions' could be exergonic in neighboring (but distinct) microenvironments. It should be noted that bacteria that may perform 'back reactions' also have alternate energy-yielding reactions available, and their mere presence in digesters suggests, but does not demonstrate, that 'back reactions' are exergonic there.

3.5 SUMMARY

Conversion of soluble intermediates of the anaerobic decomposition of polymeric components involves a complex network of microbial populations that is to a great extent controlled by flow of electrons in the form of H_2. In digester systems it is likely that numerous interspecies H_2-transfer interactions exist, involving many different substrates and couplets. These interactions permit the overall degradation of the starting polymers completely to CH_4 and CO_2, with acetate and H_2–CO_2 as the two key methanogenic substrates. Without the interspecies H_2-transfer reactions, intermediate breakdown products would accumulate and the starting substrates would not be degraded completely to CH_4 and CO_2. These and other complex interactions no doubt serve to consolidate and stabilize the anaerobic digestion process. The goal of anaerobic digestion is to produce methane from these intermediates without accumulation of organic acids or other fermentation products that inhibit the overall fermentation by pH reduction or other mechanisms. Measurement of H_2 by using new, sensitive instruments is being considered as a method of early detection of the onset of digester failure. Elevated concentration of H_2 may indicate beginning failure of syntropic conversion of propionate and butyrate as well as interspecies transfer that prevents formation of these acids. This may permit use of control measures to reverse the inhibition before the microbial community is disturbed by accumulation of large quantities of fermentation products.

REFERENCES

1. Lorowitz, W. H. and Bryant, M. P., 'Methanogenic Stearate Enrichment Cultures,' *Abs. Amer. Soc. Microbiol. Ann. Mtg.* **148**, 1983.
2. Wolin, M. J., 'Hydrogen Transfer in Microbial Communities,' pp. 323–356 in A. T. Bull and J. H. Slater, Eds., *Microbial Interactions and Communities*. New York and London: Academic Press, 1982.
3. Boone, D. R., 'Terminal Reactions in the Anaerobic Digestion of Animal Waste,' *Appl. Environ. Microbiol.* **43**:57–64, 1982.
4. Mackie, R. I. and Bryant, M. P., 'Metabolic Activity of Fatty Acid-Oxidizing Bacteria and the Contribution of Acetate, Propionate, Butyrate and CO_2 to Methanogenesis in Cattle Waste at 40 and 60 °C,' *Appl. Environ. Microbiol.* **41**:1363–1373, 1981.
5. Boone, D. R., 'Propionate Exchange Reactions in Methanogenic Ecosystems,' *Appl. Environ. Microbiol.* **48**:863–864, 1984.
6. Bryant, M. P., Wolin, E. A., Wolin, M. J. and Wolfe, R. S., '*Methanobacillus omelianskii*, a Symbiotic Association of Two Species of Bacteria,' *Arch. Mikrobiol.* **59**:20–31, 1967.
7. Iannotti, E. L., Kafkewitz, P., Wolin, M. J. and Bryant, M. P., 'Glucose Fermentation Products of *Ruminococcus albus* Grown in Continuous Culture with *Vibrio succinogenes*: Changes Caused by Interspecies Transfer of H_2,' *J. Bacteriol.* **114**:1231–1240, 1973.
8. Hungate, R. E., *The Rumen and Its Microbes*. New York and London: Academic Press, 1966.
9. Scheifinger, C. C., Linehan, B. and Wolin, M. J., 'H_2 Production by *Selenomonas ruminantium* in the Absence and Presence of Methanogenic Bacteria,' *Appl. Microbiol.* **29**:480–483, 1975.
10. Reddy, C. A., Bryant, M. P. and Wolin, M. J., 'Characteristics of S Organism Isolated From *Methanobacillus omelianskii*,' *J. Bacteriol.* **109**:539–545, 1972.
11. Wolin, M. J., 'Metabolic Interactions Among Intestinal Microorganisms,' *Am. J. Clin. Nutr.* **27**:1320–1328, 1974.
12. Chen, M. and Wolin, M. J., 'Influence of CH_4 Production by *Methanobacterium ruminantium* on the Fermentation of Glucose and Lactate by *Selenomonas ruminantium*,' *Appl. Environ. Microbiol.* **34**:756–759, 1977.
13. Chung, K.-T., 'Inhibitory Effects of H_2 on Growth of *Clostridium cellobioparum*,' *Appl. Environ. Microbiol.* **31**:342–348, 1976.
14. Weimer, P. J. and Zeikus, J. G., 'Fermentation of Cellulose and Cellobiose by *Clostridium thermocellum* in the Absence and Presence of *Methanobacterium thermoautotrophicum*,' *Appl. Environ. Microbiol.* **33**:289–297, 1977.
15. McCarty, P. L., 'Anaerobic Waste Treatment Fundamentals, I, Chemistry and Microbiology,' *Public Works* **95**:9, 107, 1964.
16. Kaspar, H. F. and Wührmann, K., 'Kinetic Parameters and Relative Turnovers of Some Important Catabolic Reactions in Digesting Sludge,' *Appl. Environ. Microbiol.* **36**:1–7, 1978.
17. Smith, P. H., 'Studies of Methanogenic Bacteria in Sludge,' EPA-600/2-80-093. Research Report, 1980.

18. McInerney, M. J., Bryant, M. P. and Pfennig, N., 'Anaerobic Bacterium That Degrades Fatty Acids in Syntrophic Association With Methanogens,' *Arch. Microbiol.* **122**:129–135, 1979.
19. McInerney, M. J., Bryant, M. P., Hespell, R. B. and Costerton, J. W., '*Syntrophomonas wolfei* gen. nov. sp. nov., An Anaerobic, Syntrophic, Fatty Acid-Oxidizing Bacterium,' *Appl. Environ. Microbiol.* **41**:1029–1039, 1981.
20. Beaty, P. S. and McInerney, M. J., 'Isolation of *Syntrophomonas wolfei* on Crotonate,' *Abst. Ann. Meeting Am. Soc. Microbiol.* **I132**, 1986.
21. Chynoweth, D. P. and Mah, R. A., 'Volatile Acids Formation in Sludge Digestion,' *Adv. Chem. Ser.* **105**:41–54, 1971.
22. Boone, D. R. and Bryant, M. P., 'Propionate-Degrading Bacterium, *Syntrophobacter wolinii* sp. nov. gen. nov., from Methanogenic Ecosystems,' *Appl. Environ. Microbiol.* **40**:626–632, 1981.
23. McCarty, P. L., 'Energetics and Kinetics of Anaerobic Treatment,' *Am. Chem. Soc. Adv. Chem. Ser.* **105**:91–107, 1971.
24. Henson, J. M. and Smith, P. H., 'Isolation of a Butyrate-Utilizing Bacterium in Co-Culture With *Methanobacterium thermoautotrophicum* from a Thermophilic Digester,' *Appl. Environ. Microbiol.* **49**:1461–1466, 1985.
25. Zinder, S. H. and Koch, M., 'Non-Aceticlastic Methanogenesis from Acetate: Acetate Oxidation by a Thermophilic Syntrophic Coculture,' *Arch. Microbiol.* **138**:263–272, 1984.
26. Ben-Bassat, A., Lamed, R. and Zeikus, J. G., 'Ethanol Production by Thermophilic Bacteria: Metabolic Control of End Product Formation in *Thermoanaerobium brockii*,' *J. Bacteriol.* **146**:192–199, 1981.
27. Bryant, M. P., Campbell, L., Reddy, C. A. and Crabill, M. R., 'Growth of Desulfovibrio in Lactate or Ethanol Media Low in Sulfate in Association with H_2-Utilizing Methanogenic Bacteria,' *Appl. Environ. Microbiol.* **33**:1162–1169, 1977.
28. Prins, R. A. and Clarke, R. T. J., 'Microbial Ecology of the Rumen,' pp. 179–204 in Y. Ruckebush and P. Thievend, Eds., *Digestive Physiology and Metabolism in Ruminants*. Lancaster, UK: M.T.P. Press, 1980.
29. Prins, R. A. and Lankhorst, A., 'Synthesis of Acetate From CO_2 in the Cecum of Some Rodents,' *FEBS Lett.* **1**:255–258, 1977.
30. Levy, P. F., Barnard, G. W., Garcia-Martinez, D. V., Sanderson, J. E. and Wise, D. L., 'Organic Acid Production From CO_2–H_2 by Mixed-Culture Anaerobes,' *Biotechnol. Bioeng.* **23**:2293–2306, 1981.
31. Goldberg, I. and Cooney, C. L., 'Formation of Short-Chain Fatty Acids From H_2 and CO_2 by a Mixed Culture of Bacteria,' *Appl. Environ. Microbiol.* **41**:148–154, 1981.
32. Lettinga, G., van der Geest, A. T., Hobma, S. and van der Laan, J., 'Anaerobic Treatment of Methanolic Wastes,' *Water Res.* **13**:725–737, 1979.
33. Lettinga, G., de Zeeuw, W. and Ouborg, E., 'Anaerobic Treatment of Wastes Containing Methanol and Higher Alcohols,' *Water Res.* **15**:171–218, 1981.
34. Sharak-Genthner, B. R., Davis, C. L. and Bryant, M. P., 'Features of Rumen and Sewage Sludge Strains of *Eubacterium limosum*, a Methanol- and H_2–CO_2-Utilizing Species,' *Appl. Environ. Microbiol.* **42**:12–19, 1981.

35. Zeikus, J. G., Lynd, L. H., Thompson, T. E., Krzycki, J. A., Weimer, P. J. and Hegge, P. W., 'Isolation and Characterization of a New, Molylotropic, Acidogenic Anaerobe, the Marburg Strain,' *Curr. Microbiol.* **3**:381–386, 1980.
36. Lorowitz, W. H. and Bryant, M. P., '*Peptostreptococcus productus* Produces Acetate and Succinate from Carbon Monoxide,' *Abstracts, Symp. on Microbial Ecol.*, 1983.
37. Laanbroek, H. J., Abee, T. and Voogd, I. L., 'Alcohol Conversions by *Desulfobulbus propionicus* Lindhorst in the Presence and Absence of Sulfate and Hydrogen,' *Arch. Microbiol.* **133**:178–184, 1982.
38. Boone, D. R., 'Mixed-Culture Fermentor for Simulating Methanogenic Digestors,' *Appl. Environ. Microbiol.* **48**:122–126, 1984.
39. Boone, D. R., 'Thermodynamics of Catabolic Reactions in the Anaerobic Digestor,' pp. 85–96 of A. A. Antonopoulos, Ed., *Biotechnological Advances in Processing Municipal Wastes for Fuels and Chemicals.* Argonne, IL: Argonne National Laboratory, 1985.
40. Balch, W. E., Schoberth, S., Tanner, R. S. and Wolfe, R. S., '*Acetobacterium*, a New Genus of Hydrogen-Oxidizing, Carbon Dioxide-Reducing, Anaerobic Bacteria,' *Int. J. Syst. Bacteriol.* **27**:355–361, 1977.
41. Fontaine, F. E., Peterson, W. H., McCoy, E., Johnson, M. J. and Ritter, G. J., 'A New Type of Glucose Fermentation by *Clostridium thermoaceticum* n. sp.,' *J. Bacteriol.* **43**:701–715, 1942.
42. Kerby, R. and Zeikus, J. G., 'Growth of *Clostridium thermoaceticum* on H_2–CO_2 or CO as an Energy Source,' *Curr. Microbiol.* **8**:27–30, 1983.
43. Braun, M. and Gottschalk, G., '*Acetobacterium wieringae* sp.-nov., A New Species Producing Acetic Acid from Molecular Hydrogen and Carbon Dioxide,' *Zbl. Bakt. Mikrobiol. Hyg. I, C-Allg.* **3**:368–376, 1982.
44. Weiringa, K. T., 'The Formation of Acetic Acid From Carbon Dioxide and Hydrogen by Anaerobic Spore-Forming Bacteria,' *Antonie van Leeuwenhoek J. Microbiol. Serol.* **6**:251–262, 1941.
45. Braun, M., Mayer, F. and Gottschalk, G., '*Clostridium aceticum* (Weiringa), a Microorganism Producing Acetic Acid From Molecular Hydrogen and Carbon Dioxide,' *Arch. Microbiol.* **128**:288–293, 1981.
46. Andreesen, J. R., Gottschalk, G. and Schlegel, H. G., '*Clostridium formicoaceticum* nov. sp., Isolation, Description, and Distinction from *C. aceticum* and *C. thermoaceticum*,' *Arch. Microbiol.* **72**:154–174, 1970.
47. Leigh, J. A., Mayer, F. and Wolfe, R. S., '*Acetogenium kivui*, a New Thermophilic Hydrogen-Oxidizing, Acetogenic Bacterium,' *Arch. Microbiol.* **129**:275–280, 1981.

4

Methanogenic Bacteria

THOMAS FERGUSON* and ROBERT MAH
School of Public Health, University of California, Los Angeles, California, USA

4.1 INTRODUCTION

The methanogenic (methane-producing) bacteria carry out the terminal step in the formation of biogas from the anaerobic decomposition of biomass. Methane is the final product of mineralizing the organic materials in digesters and most anaerobic, freshwater habitats (Fig. 4.1). Most of the chemical energy in the starting materials (substrates) actually ends up in the methane released by these anaerobic bacteria. In direct contrast, aerobic bacterial metabolism releases most of the chemical energy in the starting substrates by oxidizing them to carbon dioxide and water; bacterial cells are also produced. Anaerobic processes not only form an energy-rich product but also make less cell material and are consequently doubly useful in the degradation of biomass.

The biology of the methanogenic bacteria has been extensively reviewed during the past decade.[1–7] In this chapter, the methanogens are discussed in terms of anaerobic digestion, and the rapidly growing list of methanogens is updated.

4.2 ECOLOGY OF THE METHANOGENS

Scientific interest in the gases produced by natural decomposition of organic matter was first reported in the sixteenth century by Robert Boyle and Stephen Hale,[8] who noted that an inflammable gas was released by disturbing the sediment of natural streams and lakes. This 'inflammable air' was associated with decomposing organic material in the sediments. These early observations are easily duplicated today by inverting a large

*Present address: School of Medicine, University of California, Davis, California, USA

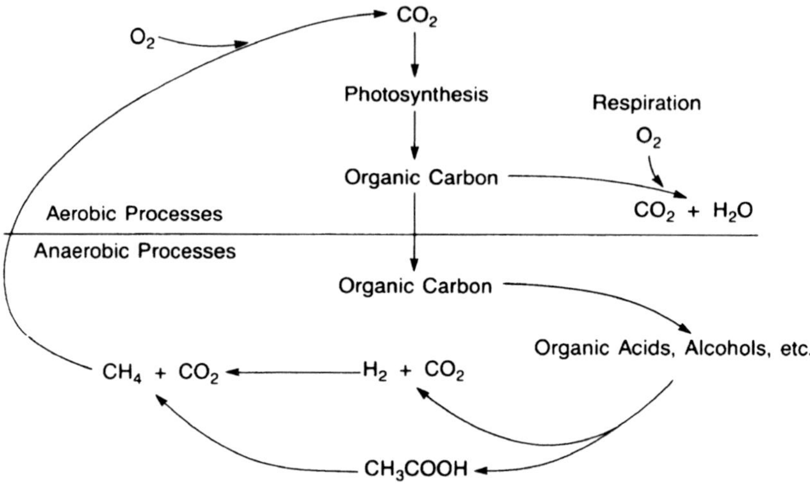

FIG. 4.1. Cycling of carbon.

funnel over nutrient-rich sediments, and collecting the gas released when they are agitated; the entrapped gas may be ignited. Early observations by Soehngen[9] led to the conclusion that fermentation intermediates produced from cellulose, and not cellulose itself, were the substrates for methanogenesis. However, culture techniques for isolating strictly anaerobic bacteria were not available until the Hungate roll-tube technique[10,11] was developed. Hence, verification of this hypothesis about the precursor substrates for methane was delayed until pure (axenic) cultures of methanogens were finally isolated. Now it is commonly accepted that methanogens are limited to a few substrates including H_2–CO_2, formate, methanol, acetate, and mono-, di- and trimethylamines. Some strains can oxidize higher alcohols[12] and others can produce methane from methylated compounds such as dimethylselenide[13] or dimethylsulfide.[14] These compounds are probably not important in anaerobic digestion. In a typical anaerobic digester, the two major methanogenic precursors are acetate and H_2–CO_2. Approximately 70% of the digestor methane comes from acetate and the remainder from CO_2 reduction to CH_4.[15,16] This relationship holds for both mesophilic (30 to 40 °C) or thermophilic (50 to 65 °C) anaerobic digestion processes. Since the methanogenic bacteria do not produce acetate or H_2 as major products, these compounds are obviously produced by nonmethanogenic bacteria. However, the relationship between the methanogens and nonmethanogens is not a simple food chain effect in which nonmethanogens produce low-molecular-weight products used as substrates by the methanogens. The interrelationship is a complex and intimate one expressed at the physiological level and upon which both methanogens and nonmethanogens are obligately linked together (Chapter 3). By oxidizing H_2, the methanogens remove it and keep its concentration low enough for H_2-sensitive chemo-organotrophs to grow and metabolize compounds they would otherwise be incapable of using. Thus, the methanogens not only obtain energy by oxidizing H_2, but also operate an 'electron sink' for H_2-sensitive chemo-organotrophs.[2,3,17] The interaction of H_2-producing and H_2-oxidizing species in this process of interspecies H_2 transfer[2,3] (Chapter 3) permits mechanisms by which

(a) substrates not readily metabolized as carbon and energy sources by pure cultures of anaerobic bacteria may be metabolized by mixed cultures in anaerobic environments;
(b) more complete bioconversion of organic materials may occur under anaerobic conditions;
(c) methanogenic bacteria may obtain H_2, CO_2, and acetate from complex substrates such as proteins, carbohydrates, and lipids that could otherwise not serve directly as substrates for methanogenesis.

4.3 THE METHANOGENIC BACTERIA

The methanogenic bacteria are among the most fastidious of anaerobes ever grown in pure culture. This characteristic undoubtedly contributed to the inability to isolate axenic cultures by early investigators. Development of the Hungate roll-tube technique[11] and discovery of a fluorescent property related to unique cofactors found in methanogens[18] greatly facilitated isolation and work with these bacteria.[19] Methanogens also require or are stimulated by growth factors such as vitamins, unusual trace minerals (Co, Ni), fatty acids (acetate, 2-methylbutyrate) or specific cofactors (coenzyme M) unique to the methanogenic bacteria. However, a few isolates may grow chemolithotrophically.[7,20] Methanogenic bacteria were subsequently isolated from a variety of habitats including anaerobic sediments (freshwater and marine), the heartwood of trees, flooded soils, the digestive tracts and feces of man and animals, as well as extreme environments such as saturated brine ponds, thermal hot springs, thermal marine vents, and an 85 °C Icelandic hotspring.[2,7,21,22] A number of new species and genera of methanogens were recently isolated in several different laboratories. The list of axenic cultures of methanogens has grown from four at the beginning of the 1960s to hundreds in 1986, including 33 separate, recognized species. As a group they all share the physiological property of forming methane from the limited substrates already listed. H_2–CO_2 is used by nearly all methanogens, but the remainder of the substrates divide the methanogens into two subcategories: methylotrophic, using acetate, methanol, methylated amines, and other methylated compounds; or nonmethylotrophic, using H_2–CO_2 and sometimes formate. Their ultrastructure and molecular composition indicate a basic divergence from other bacteria in the makeup of the cell wall and in the basic structure of the 16S ribosomal RNA. Table 4.1 lists the species reported to date, the habitats from which they were isolated, and their patterns of substrate use. Of the compounds which may serve as methanogenic substrates (Fig. 4.2), methanol and other methylated compounds are probably of little significance in anaerobic digesters. Since approximately 70% of the digester methane originates from acetate,[15,16] this is obviously the most important precursor on which to focus. The remainder of the digester methane (approximately 30%) originates mainly from the reduction of CO_2 by H_2.

Some methanogens are autotrophic and some require complex growth factors. Findings from mixed culture studies[23–25] led early workers to use trace metal solutions in pure culture studies of methanogens. In most cases habitat-simulating media were used which also contained minerals and growth factors. Jones and Stadtman[57] discovered tungstate and selenite requirements for formate metabolism by *Methanococcus vannielii* when deteriorating tap water used in their medium was replaced by distilled water. Likewise, engineering studies[25] indicated nickel requirements for active methanogenesis, and subsequently[58] *Methanobacterium thermoautotrophicum* was shown to require nickel, cobalt, and molybdenum for optimal growth and methane production from H_2–CO_2. The requirement for nickel is unusual in biological systems. Studies[23] of optimal metal concentrations for mixed methanogenic cultures indicated extremely high levels of ferrous ion gave best growth. This stimulation may have been due to traces of nickel in the iron solution (P. L. McCarty, personal communication). In methanogens Ni is incorporated into the structure of Factor 430, a cofactor in H_2 metabolism.[59] Methanogens are commonly cultured in the presence of added trace metals to satisfy these and other unknown requirements.[60] A more detailed discussion of requirements for inorganic nutrients is presented in Chapter 7. Several

TABLE 4.1
Currently Recognized Type Species of Methanogenic Bacteria

Genus and species	*Substrates*	*Refs.*
Methanobacterium formicicum DSM 863	H_2–CO_2, formate	26
Methanobacterium thermoautotrophicum ΔH	H_2-CO_2	27
Methanobacterium bryantii M.O.H.	H_2-CO_2	28,1
Methanobacterium wolfei DSM 2970	H_2-CO_2	29
Methanobacterium uliginosum P2St	H_2-CO_2	30
Methanobacterium alcaliphilum WeN4	H_2-CO_2	31
Methanobrevibacter ruminantium M1	H_2-CO_2, formate	32
Methanobrevibacter smithii PS	H_2-CO_2, formate	1
Methanobrevibacter arboriphilicus DH1	H_2-CO_2	1, 33
Methanothermus fervidus DSM 2088	H_2-CO_2	21
Methanococcus vannielii DSM 1224	H_2-CO_2, formate	34, 35
Methanococcus voltae PS	H_2-CO_2, formate	1
Methanococcus thermolithotrophicus DSM 2095	H_2-CO_2, formate	20
Methanococcus maripaludis JJ	H_2-CO_2, formate	36
Methanococcus jannaschii JAL-1	H_2-CO_2	37
Methanococcus halophilus INMI Z-7982	Methanol, trimethylamine*	38
Methanospirillum hungatei JF1	H_2-CO_2, formate	39
Methanomicrobium mobile BP	H_2-CO_2, formate	40, 19
Methanomicrobium paynteri G-2000	H_2-CO_2	41
Methanogenium cariaci JR1	H_2-CO_2, formate	42
Methanogenium marisnigri JR1	H_2-CO_2, formate	42
Methanogenium thermophilicum CR1	H_2-CO_2, formate	19
Methanogenium aggregans MSt	H_2-CO_2, formate	44
Methanogenium bourgense MS2	H_2-CO_2, formate	45
Methanosarcina barkeri MS	H_2-CO_2, methanol, trimethylamine,* acetate	26, 46.
Methanosarcina mazei S-6	Methanol, trimethylamine,* acetate	47, 48, 35
Methanosarcina acetivorans C2A	H_2-CO_2, methanol, trimethylamine,* acetate	49
Methanosarcina thermophila TM-1	Methanol, trimethylamine,* acetate	50
Methanoplanus limicola DSM 2279	H_2-CO_2, formate	51
Methanococcoides methylutens TMA-10	Methanol, trimethylamine*	52
Methanolobus tindarius Tindari 3	Methanol, trimethylamine*	53
Methanothrix soehngenii Opfikon	Acetate	54
Methanothrix concilii GP6	Acetate	55
Methanosphaera stadmanae MCB-3	Methanol plus H_2†	56

*May also use mono- and dimethylamine.
†Requires the combination of methanol and H_2.

1. Hydrogen
   $4\ H_2 + CO_2 \rightarrow CH_4 + 2\ H_2O$

2. Acetate
   $CH_3COOH \rightarrow CH_4 + CO_2$

3. Formate
   $4\ HCOOH \rightarrow CH_4 + 3\ CO_2 + 2\ H_2O$

4. Methanol
   $4\ CH_3OH \rightarrow 3\ CH_4 + CO_2 + 2\ H_2O$

5. Trimethylamine
   $4\ (CH_3)_3N + 6\ H_2O \rightarrow 9\ CH_4 + 3\ CO_2 + 4\ NH_3$

6 Dimethylamine
   $2\ (CH_3)_2NH + 2\ H_2O \rightarrow 3\ CH_4 + CO_2 + 2\ NH_3$

7. Monomethylamine
   $4\ (CH_3)NH_2 + 2\ H_2O \rightarrow 3\ CH_4 + CO_2 + 4\ NH_3$

FIG. 4.2. Methanogenic reactions.

methanogens also exhibit organic growth factor requirements. *Methanobacterium ruminantium* requires coenzyme M, the fatty acids 2-methylbutyrate and acetate, as well as certain amino acids.[61] Acetate is required for growth by several other methanogenic isolates.[62] *Methanococcus voltae* requires the amino acids leucine and iso-leucine for growth.[63] *Methanomicrobium mobile* requires the vitamins thiamine, pyridoxine, and *p*-aminobenzoic acid for growth.[64] Because of the importance of aceticlastic (acetate splitting) methanogens in anaerobic digestion, more attention should be given to identification of specific growth requirements for this group of methanogens. Identification of growth requirements could be of major importance to the understanding and stimulation of anaerobic digestion processes. The aceticlastic reaction that produces CH_4 and CO_2 is limited to only two of the many genera of methanogenic bacteria. Acetate is cleaved by some as yet unknown reaction involving an intramolecular rearrangement of protons. The reaction is thermodynamically low in net energy formed and hence there has been some controversy over its ability to serve as a sole energy source for growth of methanogens. The evidence is now incontrovertible that acetate by itself can support the growth of methanogens. Of the two known aceticlastic genera, *Methanosarcina* and *Methanothrix*, less is known about *Methanothrix*. *Methanosarcina*, however, has been isolated on many occasions by many different investigators. Members of this genus included both mesophilic and thermophilic representatives, but not all strains grew well in the presence of H_2. In fact, the thermophilic *Methanosarcina thermophila* TM-1 does not oxidize H_2, and the utilization of acetate is inhibited by its addition.[65] In aceticlastic populations of *Methanosarcina barkeri* 227,[66] the addition of H_2 only temporarily inhibits acetate utilization since *Methanosarcina barkeri* can use the added H_2. In *Methanosarcina mazei* S-6, H_2–CO_2 is used more slowly than acetate or methanol. In this case, addition of H_2–CO_2 to an aceticlastic population inhibits acetate utilization for a longer period of time (than *Methanosarcina barkeri* 227). Likewise, methanol[67] or trimethylamine[68] may also regulate acetate utilization in mesophilic strains of *Methanosarcina*. The role of substrate regulation at the level of methanogenesis remains to be further examined and is obviously important to the overall understanding of regulation in mixed culture systems, particularly in anaerobic digesters. *Methanothrix soehngenii* (formerly *Methanobacterium soehngenii*) is the only methanogen to grow exclusively on acetate. It apparently does not grow on any of the other substrates listed in Table 4.1. Its high affinity for acetate (K_m = 0·46 mM) and its long generation time may make this an important strain in digesters or anaerobic filter systems with long retention times.[69] Isolation and study of similar methanogens from other digesters will help to evaluate their role in anaerobic systems. Early study of the biochemical mechanism for cleavage of acetate was complicated by the necessity to exclude oxygen from cell-free extracts. In spite of this, the first studies by H. A. Barker and his coworkers[9,70–72] on *Methanosarcina barkeri* were successful, and the findings were later confirmed by other groups. Acetate is cleaved by decarboxylation, the methyl group being reduced to methane with its hydrogens intact (Fig. 4.2, eqn (2)). Very little of the methyl group of acetate is converted to CO_2; conversely, almost none of the carboxyl group of acetate is reduced to methane by axenic strains of *Methanosarcina*[65,67] or *Methanothrix soehngenii*.[69,54] Coenzyme M is probably involved in the final reduction step of methane production from acetate,[5,73] but little is known about the actual steps in this reaction.

Formic acid is produced in significant quantities by many fermentative organisms. Many methanogens from anaerobic digesters metabolize formic acid as an energy source. Formate may be oxidized by formate dehydrogenase to CO_2 and protons, with the electrons passed to Factor

420. This is the same electron carrier which is reduced by hydrogenase, and these electrons may be used for the reduction of CO_2.[74] Wolfe and his colleagues at the University of Illinois, and others,[4,5] isolated and characterized a number of interesting enzymes and cofactors from methanogenic bacteria. Coenzyme M (CoM), the first cofactor unique to methanogens,[75,76] was identified as thiolethanesulfonic acid ($HS\text{-}SCH_2\text{-}CH_2SO_3^-$) and serves as the terminal methyl carrier in methane formation.[5,61,76–78] This final reaction was catalyzed by methyl-CoM methylreductase.[73] Another cofactor known as F_{420} served as an electron carrier to couple hydrogenase to the metabolism of H_2 during CH_4 formation in *Methanobacterium ruminantium.*[79] These two cofactors are the first specific compounds for which a role may be hypothesized during methanogenesis from H_2 and CO_2. It is likely that other methanogenic substrates may feed directly or indirectly into pathways leading to the participation of these same cofactors.

The genetics of methane-producing bacteria is in its infancy. The results of preliminary studies have attracted a great deal of interest because of the potential construction of methanogens capable of more effective substrate use and rapid increases in biogas generation. Deoxyribonucleic acid (DNA) isolated from *Methanobrevibacter arbophilicus* was successfully introduced and partially expressed using vector plasmids in *Escherichia coli.*[80] Plasmids have been tentatively identified in at least one methanogenic coccus.[81] These small extrachromosomal genetic carriers are of basic importance in the manipulation and transfer of information from one bacterial cell to another. These results, together with the report of bromoethanesulfonate resistant *Methanosarcina* mutants[82] and the increased focus on developing more mutants among the methanogens, indicate the rapid progress being made in this field. Much groundwork still needs to be done, including isolation of new methanogens and basic studies on the biology and physiology of these organisms. This type of information is the foundation for further genetic and molecular studies. The methanogenic bacteria, together with *Halobacterium*, *Thermoplasma*, and *Sulfolobus*, form a distinct biological group called the *Archaeobacteria.*[83] This assignment is based on the concept that certain basic macromolecules are chronometric and can be used to indicate evolutionary relationships.[83] An examination of the 16S ribosomal RNA catalogues compiled for members of this group indicates that these genera are more closely related to each other than to other prokaryotic or eukaryotic organisms.[84] This concept is supported by other significant differences in cell wall structure and chemistry[85] as well as composition of cell membrane lipids.[86] The methanogenic members of the *Archaeobacteria* contain the largest number of species and genera. The evolutionary biochemistry and molecular biology of this latter group and its relation to other members of the *Archaeobacteria*, other prokaryotes, and eukaryotes are being examined in several laboratories.

4.4 SUMMARY

When one examines the characteristics of the methanogenic bacteria in pure culture, it is apparent that most axenic cultures are primarily H_2–CO_2 users. Yet in freshwater sediments and anaerobic digesters acetate is the predominant precursor of methane. *Methanothrix soehngenii* may not be axenic, but another *Methanothrix* species has been isolated.[55] Problems in maintaining strictly anaerobic conditions for extended periods, the lack of selective media, and the rapid rates of growth of H_2-oxidizing methanogens have made it difficult to isolate additional aceticlastic methanogens. Most research regarding the microbiology of anaerobic digestion has focused on the characterization of H_2-oxidizing methanogenic bacteria because of their relative ease of culturing. This research has produced significant breakthroughs regarding the nutrition and biochemistry of methanogenic bacteria in general. Attempts to study populations of methanogens have been complicated by the general lack of information on the extent of this subgroup, the requirement for maintaining reduced O_2 concentrations for extended time periods, unusual and unknown growth factors, and the expenditure of time, effort, and equipment necessary for successful isolation and maintenance of members of this unique subgroup of *Archaeobacteria*, the methanogens. Although aceticlastic methanogens are the most significant direct producers of methane in anaerobic digesters, only a few species have been cultured and characterized. Most extant axenic cultures belong to the genus *Methanosarcina*; most utilize acetate, methanol, methylamines, and H_2–CO_2 for growth and methanogenesis. However, not all aceticlastic isolates from anaerobic digesters utilize H_2–CO_2. Other methylotrophic methanogens from saline and hypersaline environments appear to be physiologically related to *Methanosarcina* but lack the ability to use both acetate and H_2–CO_2.

Since acetate accounts for 65–85% of the methane produced during the digestion of sewage sludge or animal wastes, study of the aceticlastic methanogens should be given high priority. It is unlikely that significant progress toward the understanding of anaerobic digestion will be made without first examining and characterizing this important metabolic group. Study of the regulation of substrate metabolism by *Methanosarcina*, or other as yet unknown aceticlastic methanogens, may be important to understanding the regulation of acetate metabolism under the prevailing digester conditions. Aceticlastic methanogens from anaerobic digesters may be stimulated by the addition of sludge supernatant fluid or other complex nutrients containing unknown growth factors. Further information on such requirements may have value in stimulating methanogenesis in mixed anaerobic digester systems. Any further study of the methanogenic bacteria must be integrated with the study of the fermentative and acetogenic bacteria from the same digesters. Knowledge of the physiological interactions of known cocultures may lead to significant breakthroughs in understanding and manipulating the anaerobic digestion of biomass.

REFERENCES

1. Balch, W. E., Fox, G. E., Magrum, L. J., Woese, C. R. and Wolfe, R. S., 'Methanogens: Reevaluation of a Unique Biological Group,' *Microbiol. Rev.* **43**:260–296, 1979.
2. Mah, R. A., Ward, D. M., Baresi, L. and Glass, T. L., 'Biogenesis of Methane,' *Ann. Rev. Microbiol.* **31**:309–341, 1977.
3. Mah, R. A., 'Methanogenesis and Methanogenic Partnerships,' *Philosoph. Trans. Roy. Soc., London*, **297**:599–616, 1982.
4. Wolfe, R. S., 'Methanogenesis,' *Microbiol. Biochem.* **21**:270–295, 1979.
5. Wolfe, R. S., 'Respiration in Methanogenic Bacteria,' Chapter 6, pp. 161–186 in C. J. Knowles, Ed., *Diversity of Bacterial Respiratory Systems, Vol. 1.* Boca Raton, Fla.: CRC Press, 1980.
6. Zehnder, A. J. B., 'Ecology of Methane Formation,' Chapter 13, pp. 349–376 in R. Mitchell, Ed., *Water Pollution Microbiology, Vol. 2.* New York: John Wiley & Sons, 1978.
7. Zeikus, J. G., 'The Biology of Methanogenic Bacteria,' *Bacteriol. Rev.* **41**:514–541, 1977.
8. Pine, M. J., 'The Methane Fermentations,' *Adv. Chem. Ser.* **105**:1–11, 1972.
9. Barker, H. A., 'Biological Formation of Methane,' Chapter 1 in *Bacterial Fermentations.* New York: John Wiley & Sons, 1956.
10. Hungate, R. E., *The Rumen and Its Microbes.* New York: Academic Press, 1966.
11. Hungate, R. E., 'A Roll Tube Method for Cultivation of Strict Anaerobes,' pp.117–132 in R. Norris and D. W. Ribbons, Eds., *Methods in Microbiology, Vol. 3B.* New York: Academic Press, 1969.
12. Widdel, F. and Wolfe, R. S., 'Growth of Methanogenic Bacteria in Pure

Culture with 2-propanol as Sole Hydrogen Donor,' *Abst. Ann. Meeting Am. Soc. Microbiol.* **I-126**, 1986.

13. Zehr, J. P. and Oremland, R. S., 'Anaerobic Degradation of Dimethylselenide to CH_4 and CO_2 by Sediments and a Methylotrophic Methanogen,' *Abst. Ann. Meeting Am. Soc. Microbiol.* **I-130**, 1986.
14. Oremland, R. S. and Kiene, R. P., 'Isolation and Characterization of a Methylotrophic Estuarine Methanogen Which Grows on Dimethylsulfide,' *Abst. Ann. Meeting Am. Soc. Microbiol.* **I–131**, 1986.
15. Jeris, J. S. and McCarty, P. L., 'The Biochemistry of the Methane Fermentation Using ^{14}C Tracers,' *J. Water Pollut. Control Fed.* **37**:178-192, 1965.
16. Smith, P. H. and Mah, R. A., 'Kinetics of Acetate Metabolism During Sludge Digestion,' *Appl. Microbiol.* **14**:368–371, 1966.
17. Boone, D. R., 'Terminal Reactions in the Anaerobic Digestion of Animal Waste,' *Appl. Environ. Microbiol.* **43**:57–64, 1982.
18. Miller, T. L. and Wolin, M. J., 'A Serum Vial Modification of the Hungate Technique for Cultivating Obligate Anaerobes,' *Appl. Microbiol.* **27**:985–987, 1974.
19. Cheeseman, P., Toms-Wood, A. and Wolfe, R. S., 'Isolation and Properties of a Fluorescent Compound, Factor 420, From *Methanobacterium* Strain M.o.H.,' *J. Bacteriol.* **112**:527–531, 1972.
20. Huber, H., Thomm, M., König, H., Thies, G. and Stetter, K. O., '*Methanococcus thermolithotrophicus*, A Novel Thermophilic Lithotropic Methanogen,' *Arch. Microbiol.* **132**:47–50, 1982.
21. Stetter, K. O., Thomm, M., Winter, J., Wildgruber, G., Huber, H., Zillig, W., Janecovic, D., König, H., Palm, P. and Wonderl, S., '*Methanothermus fervidus*, sp. nov., A Novel Extremely Thermophilic Methanogen from an Icelandic Hot Spring.' *Zbl. Bakt. Hyg., I. Abt. Orig.* **C2**:166–178, 1981.
22. Miller, T. L. and Wolin, M. J., 'Oxidation of Hydrogen and Reduction of Methanol to Methane is the Sole Energy Source for a Methanogen Isolated from Human Feces,' *J. Bacteriol.* **153**:1051–1055, 1983.
23. Owen, W. F., Stuckey, D. C., Healy, J. B., Young, L. Y. and McCarty, P. L., 'Bioassay for Monitoring Biochemical Methane Potential and Anaerobic Toxicity,' *Water Res.* **13**:485–492, 1983.
24. Speece, R. E. and McCarty, P. L., 'Cobalt and Iron Stimulation of Anaerobic Digestion,' *Proc. First Int. Conf. Water Pollut.*, 1962.
25. Speece, R. E., Parkin, G. F. and Galleghar, D., 'Nickel Stimulation in Anaerobic Digestion,' *Water Res.* **17**:677–683, 1983.
26. Schnellen, C. G. T. P., 'Onderzoekingen Over de Methaangisting,' Thesis, Delft, pp. 1–137, 1947.
27. Zeikus, J. G. and Wolfe, R. S., '*Methanobacterium thermoautotrophicus Sp.* N., an Anaerobic, Autotrophic, Extreme Thermophile,' *J. Bacteriol.* **109**:707–713, 1972.
28. Bryant, M. P., Wolin, E. A., Wolin, M. J. and Wolfe, R. S., '*Methanobacillus omelianskii*, a Symbiotic Association of Two Species of Bacteria,' *Arch. Mikrobiol.* **59**:20–31, 1967.
29. Winter, J., Lerp, C., Zabel, H.-P., Wildenauer, F. X., König, H. and Schindler, F., '*Methanobacterium wolfei*, Sp. Nov., a New Tungsten-

Requiring, Thermophilic, Autotrophic Methanogen,' *Syst. Appl. Microbiol.* **5**:457–466, 1985.
30. König, H., 'Isolation and Characterization of *Methanobacterium uliginosum* Sp. Nov. from a Marshy Soil,' *Can. J. Microbiol.* **30**:1477–1481, 1984.
31. Worakit, S., Boone, D. R., Mah, R. A., Abdel-Samie, M.-E. and El-Halwagi, M. M., '*Methanobacterium alcaliphilum* Sp. Nov., an H_2-Utilizing Methanogen which Grows at High pH Values,' *Int. J. Syst. Bacteriol.* **36**:380–382, 1986.
32. Smith, P. H. and Hungate, R. E., 'Isolation and characterization of *Methanobacterium ruminantium* N. Sp.,' *J. Bacteriol.* **75**:713–718, 1958.
33. Zeikus, J. G. and Henning, D. L., '*Methanobacterium arbophilicum* Sp. Nov.; an Obligate Anaerobe Isolated from Wetwood of Living Trees,' *Antonie van Leeuwenhoek J. Microbiol. Serol.* **41**:543–552, 1975.
34. Stadtman, T. C. and Barker, H. A., 'Studies on the Methane Fermentation. A New Formate-Decomposing Bacterium,' *J. Bacteriol.* **62**:269–280, 1951.
35. Mah, R. A. and Kuhn, D. A., 'Transfer of the Type Species of the Genus *Methanococcus* to the Genus *Methanosarcina*, Naming it *Methanosarcina mazei* (Barker 1936) Comb. Nov. et Emend. and Conservation of the Genus *Methanococcus* (Approved Lists 1980) with *Methanococcus vannielii* (Approved Lists 1980) as the Type Species. Request for an Opinion,' *Int. J. Syst. Bacteriol.* **34**:263–265, 1984.
36. Jones, W. J., Paynter, M. J. B. and Gupta, R., 'Characterization of *Methanococcus maripaludis* Sp. Nov., a New Methanogen Isolated from Salt Marsh Sediment,' *Arch. Microbiol.* **135**:91–97, 1983.
37. Jones, W. J., Leigh, J. A., Mayer, F., Woese, C. R. and Wolfe, R. S., '*Methanococcus jannaschii* Sp. Nov., an Extremely Thermophilic Methanogen from a Submarine Hydrothermal Vent,' *Arch. Microbiol.* **136**:254–261, 1983.
38. Zhilina, T. N., 'A New Obligate Halophilic Methane-Producing Bacterium,' *Mikrobiologiya* **52**:375–382, 1983.
39. Ferry, J. G., Smith, P. H. and Wolfe, R. S., '*Methanospirillum*, a New Genus of Methanogenic Bacteria, and Characterization of *Methanospirillum hungatii* Sp. Nov.,' *Int. J. Syst. Bacteriol.* **24**:465–469, 1974.
40. Paynter, M. J. B. and Hungate, R. E., 'Characterization of *Methanobacterium mobilis*, Sp. N., Isolated from the Bovine Rumen,' *J. Bacteriol.* **95**:1943–1951, 1968.
41. Rivard, C. J., Henson, J. M., Thomas, M. V. and Smith, P. H., 'Isolation and characterization of *Methanomicrobium paynteri* Sp. Nov., a Mesophilic Methanogen Isolated from Marine Sediments,' *Appl. Environ. Microbiol.* **46**:484–490, 1983.
42. Romesser, J. A., Wolfe, R. S., Mayer, F., Speiss, E. and Walther-Mauruschat, A., '*Methanogenium*, a New Genus of Marine Methanogenic Bacteria, and Characterization of *Methanogenium cariaci* Sp. Nov. and *Methanogenium marisnigri* Sp. Nov.,' *Arch. Microbiol.* **121**:147–153, 1979.
43. Rivard, C. J. and Smith, P. H., 'Isolation and Characterization of a Thermophilic Marine Methanogenic Bacterium, *Methanogenium thermophilicum*, Sp. Nov.,' *Int. J. Syst. Bacteriol.* **32**:430–436, 1982.
44. Ollivier, B. M., Mah, R. A., Garcia, J. L. and Robinson, R., 'Isolation and Characterization of *Methanogenium aggregans* Sp. Nov.,' *Int. J. Syst. Bacteriol.* **35**:127–130, 1985.

45. Ollivier, B. M., Mah, R. A., Garcia, J. L., Robinson, R. and Boone, D. R., 'Isolation and Characterization of *Methanogenium bourgense* Sp. Nov.,' *Int. J. Syst. Bacteriol.* **36**:297–301, 1986.
46. Mah, R. A. and Kuhn, D. A., 'Rejection of the Type Species *Methanosarcina methanica* (Approved Lists 1980), Conservation of the Genus *Methanosarcina* with *Methanosarcina barkeri* (Approved Lists 1980) as the Type Species, and Emendation of the Genus *Methanosarcina*,' *Int. J. Syst. Bacteriol.* **34**:266–267, 1984.
47. Barker, H. A., 'Studies upon the Methane-Producing Bacteria,' *Arch. Mikrobiol.* **7**:420–438, 1936.
48. Mah, R. A., 'Isolation and Characterization of *Methanococcus mazei*,' *Curr. Microbiol.* **3**:321–326, 1980.
49. Sowers, K. R., Baron, S. F. and Ferry, J. G., '*Methanosarcina acetivorans* Sp. Nov., an Acetotrophic Methane-Producing Bacterium Isolated from Marine Sediments,' *Appl. Environ. Microbiol.* **47**:971–978, 1984.
50. Zinder, S. H., Sowers, K. R. and Ferry, J. G., '*Methanosarcina thermophila* Sp. Nov., a Thermophilic, Acetotrophic, Methane-Producing Bacterium,' *Int. J. Syst. Bacteriol.* **35**:522–523, 1985.
51. Wildgruber, G., Thomm, M., König, H., Ober, K., Ricchiuto, T. and Stetter, K. O., '*Methanoplanus limicola*, a Plate-Shaped Methanogen Representing a Novel Family, the *Methanoplanaceae*,' *Arch. Microbiol.* **132**:31–36, 1982.
52. Sowers, K. R. and Ferry, J. G., 'Isolation and Characterization of a Methylotrophic Marine Methanogen, *Methanococcoides methylutens* Gen. Nov., Sp. Nov.,' *Appl. Environ. Microbiol.* **45**:684–690, 1983.
53. König, H. and Stetter, K. O., 'Isolation and Characterization of *Methanolobus tindarius*, Sp. Nov., a Coccoid Methanogen Growing Only on Methanol and Methylamines,' *Zentralbl. Bakteriol. Parasitenkd. Infektionskr. Hyg. Abt. I Orig.* **C3**:478-490, 1982.
54. Huser, B. A., Wuhrmann, K. and Zehnder, A. J. B., '*Methanothrix soehgenii* Gen. Nov. Sp. Nov., A New Acetotrophic Non-Hydrogen Oxidizing Methane Bacterium,' *Arch. Microbiol.* **132**:1–9, 1982.
55. Patel, G. B., 'Characterization and Nutritional Properties of *Methanothrix concilii* Sp. Nov., a Mesophilic, Aceticlastic Methanogen,' *Can. J. Microbiol.* **30**:1383–1396.
56. Miller, T. L. and Wolin, M. J., '*Methanosphaera stadtmaniae* Gen. Nov., Sp. Nov.: a Species that Forms Methane by Reducing Methanol with Hydrogen,' *Arch. Microbiol.* **141**:116–122, 1985.
57. Jones, J. B. and Stadtman, T. C., '*Methanococcus vannielii*: Culture and Effects of Selenium and Tungsten on Growth,' *J. Bacteriol.* **130**:1404–1406, 1977.
58. Schonheit, P., Moll, J. and Thauer, R. K., 'Nickel, Cobalt and Molybdenum Requirement for Growth of *Methanobacterium thermoautotrophicum*,' *Arch. Microbiol.* **123**:105–107, 1979.
59. Diekert, G., Klee, B. and Thauer, R. K., 'Nickel, A Component of Factor 430 From *Methanobacterium thermoautotrophicum*,' *Arch. Microbiol.* **124**: 103–106, 1980.
60. Mah, R. A. and Smith, M. R., 'The Methanogenic Bacteria,' Chapter 76, pp. 948–977 in M. P. Starr, H. Stolp, H. G. Truper, A. Balows and H. G.

Schlegel, Eds., *The Procaryotes*. Berlin and Heidelberg: Springer Verlag, 1981.

61. Taylor, C. D., McBride, B. C., Wolfe, R. S. and Bryant, M. P., 'Coenzyme M, Essential for Growth of a Rumen Strain of *Methanobacterium ruminantium*,' *J. Bacteriol.* **120**:974–975, 1974.
62. Taylor, G. T., 'The Methanogenic Bacteria,' *Prog. Ind. Microbiol.* **16**:231–329, 1982.
63. Whitman, W. B., Ankwanda, E. and Wolfe, R. S., 'Nutrition and Carbon Metabolism of *Methanococcus voltae*,' *J. Bacteriol.* **149**:852–863, 1982.
64. Tanner, R. S. and Wolfe, R. S., 'Nutrient requirements of *Methanomicrobium mobile*,' *Abstr. Ann. Meeting Am. Soc. Microbiol.* **I-90**, 1982.
65. Zinder, S. H. and Mah, R. A., 'Isolation and Characterization of a Thermophilic Strain of *Methanosarcina* Unable To Use H_2–CO_2 for Methanogenesis,' *Appl. Environ. Microbiol.* **39**:996–1008, 1979.
66. Ferguson, T. J. and Mah, R. A., 'Effect of H_2–CO_2 on Methanogenesis From Acetate or Methanol by *Methanosarcina* sp.,' *Appl. Environ. Microbiol.* **45**:348–355, 1983.
67. Smith, M. R. and Mah, R. A., 'Growth and Methanogenesis by *Methanosarcina* Strain 227 on Acetate and Methanol,' *Appl. Environ. Microbiol.* **36**:870–879, 1978.
68. Blaut, M. and Gottschalk, G., 'Effect of Trimethylamine on Acetate Utilization by *Methanosarcina barkeri*,' *Arch. Microbiol.* **133**:230–235, 1982.
69. Zehnder, A. J. B., Huser, B. A., Brock, T. D. and Wuhrmann, K., 'Characterization of an Acetate-Decarboxylating, Non-Hydrogen Oxidizing Methane Bacterium,' *Arch. Microbiol.* **124**:1–11, 1980.
70. Barker, H. A., 'Studies Upon the Methane-Producing Bacteria,' *Arch. Mikrobiol.* **7**:420–438, 1936.
71. Stadtman, T. L. and Barker, H. A., 'Studies on the Methane Fermentation. IX. The Origin of Methane in the Acetate and Methanol Fermentation of Methanosarcina.' *J. Bacteriol.* **61**:81–86, 1951.
72. Stadtman, T. L., 'Methane Fermentation,' *Ann. Rev. Microbiol.* **21**:121–142, 1967.
73. Gunsalus, R. P. and Wolfe, R. S., 'ATP Activation of Coenzyme M in *Methanobacterium ruminantium*,' *J. Bacteriol.* **135**:851–857, 1978.
74. Ferguson, T. J. and Mah, R. A., 'Isolation and Characterization of an H_2-Oxidizing Thermophilic Methanogen,' *Appl. Environ. Microbiol.* **45**:265–274, 1983.
75. McBride, B. C. and Wolfe, R. S., 'A New Coenzyme of Methyl Transfer, Coenzyme M,' *Biochemistry* **10**:2317–2324, 1971.
76. Balch, W. E. and Wolfe, R. S., 'Specificity and Biological Distribution of Coenzyme M (2-Mercaptoethanesulfonic Acid),' *J. Bacteriol.* **137**:256–263, 1976.
77. Balch, W. E. and Wolfe, R. S., 'New Approach to the Cultivation of Methanogenic Bacteria: 2-Mercaptoethane-Sulfonic Acid Dependent Growth of *Methanobacterium ruminantium* in a Pressurized Atmosphere,' *Appl. Microbiol.* **137**:256–263, 1976.
78. Taylor, C. D. and Wolfe, R. S., 'Structure and Methylation of Coenzyme M (HS-CH_2CHSO_3),' *J. Biol. Chem.* **10**:4879-4885, 1974.

79. Tzeng, S. F., Bryant, M. P. and Wolfe, R. S., 'Factor 420 Dependent Pyridine Nucleotide Linked Hydrogenase System of *Methanobacterium ruminantium*,' *J. Bacteriol.* **121**:154–191, 1975.
80. Bollschweiler, C. and Klein, A., 'Polypeptide Synthesis in *Escherichia coli* Directed by Cloned *Methanobrevibacter arboriphilus* DNA,' *Zbl. Bakt. Hyg. I. Abt. Orig.* **C3**:101–109, 1982.
81. Thomm, M., Altenbuchner, J. and Stetter, K. O., 'Evidence for a Plasmid in a Methanogenic Bacterium,' *J. Bacteriol.* **153**:1060–1062, 1983.
82. Smith, M. R. and Mah, R. A., '2-Bromoethanesulfonate: A Selective Agent for Isolating Resistant *Methanosarcina* Mutants,' *Curr. Microbiol.* **6**:321–326, 1981.
83. Woese, C. R., '*Archaebacteria* and Cellular Origins: An Overview,' *Zbl. Bakt. Hyg. I. Abt. Orig.* **C3**:1–17, 1982.
84. Fox, G. E., Pechman, K. J. and Woese, C. R., 'Comparative Cataloging of 16S Ribosomal Ribonucleic Acid: Molecular Approach to Procaryotic Systematics,' *Int. J. Syst. Bacteriol.* **135**:44–57, 1977.
85. Kandler, O. and König, H., 'Chemical Composition of the Peptidoglycan-Free Cell Walls of Methanogenic Bacteria,' *Arch. Microbiol.* **118**:141–152, 1978.
86. Tournabene, T. G., Langworthy, T. A., Holzer, G. and Oro, J., 'Squalenes, Phytanes, and Other Isoprenoids as Major Neutral Lipids of Methanogenic and Thermoacidophilic *Archaebacteria*,' *J. Molec. Evolut.* **13**:1–8, 1979.

5

Feed Composition

DOUGLAS E. JERGER

Agricultural Engineering Department, University of Florida, Gainesville, Florida, USA

and

GEORGE T. TSAO

School of Chemical Engineering, Purdue University, Lafayette, Indiana, USA

5.1 INTRODUCTION

The biomass resource base potentially includes thousands of different plant species and crop residues. Much ambiguity exists in methods of classification of biomass feeds for anaerobic digestion.

Biomass categories typically used include herbaceous, woody, freshwater, and marine. These categories, while providing information on the growth characteristics, offer little insight into the relationship between biomass composition and experimental conditions that should be considered in digester design and operation. The compositional characteristics of biomass may differ considerably between groups (i.e. woody and freshwater) but may also vary from species to species and within a species, depending on growth conditions. Substrate composition is a major factor affecting the methane yield and methane production rates from the digestion of biomass. Techniques are available to determine the compositional characteristics of biomass, while parameters such as solids, elemental, and organic analyses are important for digester design and operation.

5.2 SOLIDS

Total solids, volatile solids, and suspended solids analyses are frequently performed on digester feeds for characterization using procedures outlined in *Standard Methods.*[1] The concentration of total solids (TS) in a feed is a critical factor in materials handling and in digester design and operation. The solids concentration affects both the capital and operational costs of anaerobic digestion for operation at particular loading rates. Feeds with lower solids concentrations require larger reactors and have higher heating requirements than those with higher solids concentrations. If the feed is to be loaded on a continuous or semicontinuous basis, the solids content and feed flow rate will dictate the front-end processing and pumping requirements. Volatile solids (VS) analyses are used as a basis for estimating organic composition, determining loading rates and evaluating digester performance.

The choice of reactor design and mode of operation is highly dependent upon the physical and chemical characteristics of the digester feeds, especially the concentration of particulate associated solids. The solids content of different feeds affects the hydraulic retention time (HRT) and the feed heating requirements for anaerobic digestion. Longer solids retention times (SRTs) and microorganism retention times (MRTs) promote greater biodegradation of less biodegradable feeds, enhance digester stability, and reduce the nutrient requirements for digestion. Longer SRTs and MRTs can be promoted in several reactor designs without increasing the HRT. However, because there are significant differences in the chemical and physical properties of many biomass feeds, optimal anaerobic digester performance requires matching feed characteristics with reactor configurations.

Highly biodegradable feeds with low solids concentrations can be gasified more efficiently and economically in attached film reactors at ambient temperature. On the other hand, feeds with high particulate-associated solids concentrations are not very suitable for use in attached film reactors because of operational and mechanical problems such as clogging of the microorganism support media. Less biodegradable feeds with high particulate-associated solids concentrations and relatively low nutrient concentrations can be more effectively digested in reactors that promote long SRTs and MRTs, such as the upflow solids reactor (USR).

A single reactor configuration and set of operating conditions may not be suitable for the anaerobic digestion of all forms of biomass. Factors such as feed solids content and reactor configuration can affect the SRT, which, in turn, can be a critical parameter affecting digester nutrient requirements, biodegradability, and methane yields. A detailed discussion of these parameters is presented in Chapter 9.

5.3 ELEMENTS AND HEATING VALUE

Ultimate analyses of biomass feeds include elements (C, H, N, O, P, etc.) and determination of heating values. The methods for these analyses have been reported in *Standard Methods*, ASTM and AOAC publications.[1-3] Data from these analyses are useful for experimental design, calculation of theoretical methane yields, determination of nutrient limitations and variability, and calculation of material and energy balances across the system. Maximum theoretical methane yield determination is useful in evaluation of digester performance and provides a basis for establishing target yields for experimental work.[4] If the organic fraction of biomass is assumed to be 100% biodegradable under anaerobic conditions, the upper theoretical yield can be calculated using either heating values or compositional data. In the former method, the upper yield is equivalent to the heating value expressed as methane. In the latter, an empirical formula developed from compositional data is introduced into a stoichiometric equation (eqn (5.1)) and the conversion of organic matter to methane and CO_2 is calculated:[5]

$$C_nH_aO_b + \left(n - \frac{a}{4} - \frac{b}{2}\right) H_2O \rightarrow \left(\frac{n}{2} - \frac{a}{8} + \frac{b}{4}\right)$$

$$CO_2 + \left(\frac{n}{2} + \frac{a}{8} - \frac{b}{4}\right) CH_4 \qquad (5.1)$$

Because a fraction of the feed organic matter in all bacterial fermentations is converted to bacterial cells, theoretical yields may be corrected for cell maintenance requirements. These requirements may vary considerably for different anaerobic systems, operating conditions, and feedstocks, and has not been extensively investigated (particularly for mixed populations growing on particulate feeds). For anaerobic digestion of soluble substrates and cellulose in a continuously stirred tank reactor (CSTR), 20% of the carbohydrates and 7% of the protein are reportedly needed for cell maintenance.[6] Studies have shown greater than 90% substrate biodegradability in batch-fed systems, suggesting that the cell maintenance factor was less than 10% of the feed VS.[7] Cell maintenance factors are easy to determine for soluble feeds; however, difficulties are encountered when monitoring cell mass during digestion of particulate feeds. Microbial cell mass in microbial ecosystems in anaerobic digesters can be estimated by measurements of microbial activity such as ATP,

TABLE 5.1
Heating Values and Theoretical Methane Yields of Different Biomass and Wastes

		*Theoretical methane yield**	
	Heating value, kJ/kg VS	*SCM/kg VS added*	*SCM/wet kg added*
Aquatic			
Kelp	19 000	0·514	0·037 3
Water hyacinth	19 700	0·537	0·022 1
Herbaceous			
Bermuda grass	19 600	0·512	0·210
Kudzu	20 100	0·538	0·150
Napier grass	19 700	0·524	0·137
Cattail	19 800	0·517	0·113
Woody			
Eucalyptus	19 200	0·502	0·280
Hybrid poplar	19 800	0·527	0·440
Cottonwood	19 700	0·547	0·365
Loblolly pine	20 400	0·547	0·341
Wastes			
Primary sludge	29 400	0·793	0·022 1
Municipal solid wastes	19 100	0·512	0·326

*Not corrected for cell maintenance and synthesis; Ref. 7

dehydrogenase, DNA, and factor 420 (for methanogens). This factor could also be different in processes designed to increase cell residence times such as attached film reactors or cell recycle systems.

Upper theoretical methane yields were determined for several different biomass and waste feedstocks, including marine algae, freshwater and herbaceous plants, and crop wastes. These data, calculated on the basis of volatile solids concentration and total wet weight, are shown in Table 5.1.[7] The upper theoretical methane yields for most biomass species are similar, with a value of about 0·5 SCM/kg (8·5 SCF/lb) VS added. Assuming a cell maintenance requirement of 15%, a methane yield of about 0·47 SCM/kg (7·2 SCF/lb) VS added could be expected if all organic components were biodegradable. Theoretical methane yields, calculated on the basis of wet weight, may vary substantially among biomass species because of differences in ash and water content. As demonstrated with different kelp harvests in Table 5.2, significant

TABLE 5.2
Theoretical Methane Yields for Several Kelp Lots

Kelp lot	Stoichiometric yield*	
	SCM/kg VS added	*SCF/lb VS added*
37	0·514	8·24
42	0·444	7·11
44	0·478	7·59
46	0·403	6·45
47	0·455	7·29
49	0·536	8·58

*Assumes that the reactants are kelp + H_2O and the products are CH_4 and CO_2; data from Ref. 7.

variations in theoretical methane yields also occur within a single biomass species, depending upon growth conditions.[7] Certain wastes, such as primary sewage sludge, have a substantially higher theoretical methane yield than the biomass feeds. This may be partially attributed to the high protein and fat content of primary sludge derived from domestic sewage sources.

5.4 MEASUREMENT OF ORGANIC COMPONENTS

5.4.1 Total

Organic matter is the main source of energy for microorganisms involved in the anaerobic digestion process. If the amount of organic matter present in the feeds is known, an estimation of the methane yield can be made. Organic composition analysis of biomass can be divided into two types of measurements: those that measure the total amount of organic matter in general terms and those that are specific for individual organic compounds. Total organic matter in digester feedstocks is most frequently measured by volatile solids, chemical oxygen demand analysis, and total organic carbon analysis. All methods are limited by the difficulty of obtaining a representative sample of a heterogeneous mixture.

The volatile solids fraction of total solids analyses offers a rough approximation of the amount of organic matter present in the feed and is defined as the solids lost after ignition for 1 hour at 550 °C.[1] This method does not distinguish precisely between organic and inorganic residues because the loss on ignition includes not only organic matter but also losses due to volatilization of certain mineral salts.

Chemical oxygen demand (COD) has been cited as a better approximation of organic matter.[8] COD is a measure of the oxygen equivalent of that portion of organic matter that is susceptible to oxidation by a strong chemical oxidant. A catalyst must be added for the oxidation of straight chain aliphatic compounds (acids, alcohols); however, aromatic hydrocarbons are not oxidized to any appreciable extent. Chlorides interference in the analysis may be overcome by using a complexing agent.[1]

Total organic carbon (TOC) analysis is the most direct method of determining the amount of organic material present in the digester feed.[1] Therefore, TOC can be used to monitor processes for the digestion of biomass without dependence on oxidation states even at relatively low organic carbon concentrations. TOC is determined by combustion of organic carbon to carbon dioxide, which is then measured by means of nondispersive infrared analysis. It is necessary, however, to account for the inorganic carbon in the sample or culture volume. TOC analyses are highly sensitive and usually require modification prior to measurement of the high carbon and particulate solids concentrations in biomass feeds and digester cultures. These modifications involve reducing the sensitivity of the detector to provide accuracy in the range of 10 000–30 000 mg/liter carbon, and to develop an ample combustion technique to permit satisfactory repeatability with high solids-containing samples.[9]

5.4.2 Specific Groups

Biomass feeds are typically plant material comprised of cell wall and intracellular material. The cell wall is composed of structural carbohydrates (cellulose and hemicellulose) and lignin, while the intracellular protoplasm consists primarily of nonstructural carbohydrates (sucrose, starch) and nitrogenous compounds (proteins, nucleic acids). Each of these fractions is degraded at a different rate and to a different extent. The methane fermentation potential of biomass feeds can, therefore, be more accurately estimated using organic compositional analyses to characterize the various fractions in comparison to general techniques discussed earlier.

Organic composition analyses are often used to characterize more adequately digester feedstocks and effluents to determine biodegradability of specific components. The most frequently used technique for nonwoody biomass is the detergent fiber analyses developed by Van Soest.[10] These analyses are based on the partitioning of various organic con-

TABLE 5.3
Organic Composition* of Herbaceous Biomass

Sample	*Volatile solids, % TS*	*Neutral detergent solubles, % VS*	*Crude protein, % VS*	*Total cell wall, % VS*	*Cellulose, % cell wall*	*Hemicellulose, % cell wall*	*Lignin, % cell wall*
Napier grass	93·0	22·4	7·2	77·6	41·2	26·8	9·2
Bermuda grass	93·7	31·7	7·2	68·3	26·7	38·0	3·5
Cattail	96·0	38·9	7·2	61·1	29·0	23·6	7·8
Kudzu	93·6	39·8	13·8	60·2	31·9	16·4	10·4
Johnson grass	95·0	35·4	7·2	64·6	29·2	30·0	5·2
Water hyacinth	71·9	36·9	15·2	63·1	27·8	28·6	7·5
Kelp	73·6	67·5	30·1	32·5	12·8	13·4	6·2
Wheat straw	92·0	21·3	8·3	78·7	41·8	22·2	14·7
Corn stalks	96·2	50·1	4·8	49·1	24·4	16·8	5·8
Giant reed	94·8	31·8	6·5	68·2	32·0	27·7	6·8
Bamboo	99·8	15·8	1·8	84·2	46·3	22·2	14·9

*Organic composition analyses done by Van Soest detergent fiber method (Ref. 10). Crude protein determinations calculated by TKN X 6.25 (data taken from Refs 7 and 8).

TABLE 5.4
Organic Composition* of Woody Biomass

Sample	*Volatile solids, % TS*	*Extractables, % TS*	*% Extractive-free dry wood*		
			Lignin	*Hemicellulose*	*Cellulose*
Eucalyptus	99·2	4·44	24·77	31·84	43·39
Cottonwood	99·3	3·05	24·72	39·81	35·47
Loblolly pine	99·6	4·81	34·56	24·96	40·50
Sycamore	99·4	4·08	25·55	29·47	44·98
Hybrid poplar	99·4	5·25	25·67	32·97	41·36
Black alder	99·5	4·97	28·14	33·13	38·72

*Organic composition analyses performed by ASTM Methods as reported in Ref. 12. Data from Refs 7 and 8.

stituents by detergents. The partitioning of lignin and cellulose (acid detergent fiber) is accomplished by treatment with sulfuric acid (cellulose solubilization) or potassium permanganate (lignin solubilization). The organic composition of several herbaceous species is shown in Table 5.3. Total crude protein in biomass can be estimated from total Kjeldahl nitrogen analysis, or it can be determined directly using the Biuret, Folin-Ciocalteau, or Lowry method.[11]

A modified procedure is used for the compositional analyses of woody biomass due to its high lignin content. The method involves solubilization of the lignin with sodium chlorite leaving holocellulose residue (hemicellulose and cellulose). The hemicellulose is then solubilized with sodium hydroxide and acetic acid, leaving cellulose residue.[12] The organic composition of several woody species is shown in Table 5.4.

Biomass feeds containing high concentrations of carbohydrates (nonstructural) such as starch, sucrose, and mannitol can be analyzed by specific chromatographic techniques.[12] High-pressure liquid chromatography methods have been developed for the analysis of low-molecular-weight cellulose, and cellulose and hemicellulose hydrolysis products.[13]

Two-stage digestion of biomass feeds results in a soluble low-solids leachate from the stage-one (particulate feed) reactor being fed to the stage-two (soluble feed) reactor. Characterization of these soluble feeds has been successfully accomplished using the above techniques.[14] Another analytical method applicable to soluble feeds derived from the hydrolysis of biomass is determination of volatile fatty acids (VFA).

5.4.3 Biodegradability

Biochemical oxygen demand (BOD) is a measure of the oxygen demand produced by carbonaceous and nitrogenous compounds during biological oxidation.[1] This parameter, widely used to measure waste loadings and to evaluate the efficiency of treatment systems, has several limitations. These include, for example, the oxidation of reduced inorganic compounds, inhibition by toxic compounds, variability with different inocula, and a relatively low level of precision. The BOD test is not suitable for determining anaerobic biodegradability because of differences in electron acceptors (O_2 instead of CO_2), incubation conditions and inoculum source. Nevertheless, the test does give an indication of the biological reactivity of material under aerobic conditions, such as those required for initial hydrolysis of many organic feed components.

Studies have been made to correlate COD, BOD, TOC, and VFA analyses using wastes. BOD-to-COD ratios of between 0·2:1 and 0·5:1 have, for example, been used routinely to control wastewater treatment plant operations. COD, however, reportedly correlates better with TOC than with BOD. Although these relationships between different measures of organic matter have been studied with various wastes, they have not been established with biomass.

A rapid methane yield test has also been proposed whereby COD-to-VS ratios for various substrate constituents could be determined. Using this procedure, the methane yield potential could then reportedly be calculated by enumerating the COD-to-VS ratios of each constituent, weighted by its digestibility.[8] However, data were not presented to verify the accuracy of this procedure. Other studies, using COD data applied to certain conversion factors determined for biomass components, did not accurately predict the methane yield potential of different substrates.[15]

Both semicontinuous and batch-fed bioassays have been used to evaluate anaerobic biodegradability. The semicontinuous procedure closely simulates full-scale anaerobic operation; however, it is costly in terms of equipment, time, personnel, and facilities. Batch bioassays do not have these limitations and have been frequently used to measure substrate biodegradability.

The *in vitro* digestion assay has been used historically to measure the rumen digestibility of biomass feedstocks.[16] A short digestion period (2–3 days) is usually employed in the test to correspond to the retention time of the rumen. The test has been used to examine the biodegradability and refractory fraction for over 190 forage grasses.[17] Numerous other studies have also reported the biodegradability of various biomass types and the effects of pretreatment on the digestibility of these feeds using the *in vitro* assay.[18] However, estimation of a substrate's ultimate biodegradability may not be accurate due to the specific inoculum and the short incubation period (retention time) used for this test. In fact, in a recent study on methane fermentation biodegradability, the *in vitro* digestion technique using rumen fluid proved to be the least accurate of the methods examined.[8]

The Warburg respirometer has been widely used to measure gases evolved from the biodegradation of organic compounds; however, this method is complicated and costly. The sample size is small, making subsequent chemical analysis difficult, and it is difficult to collect samples from the gas or liquid phase during the test.[19] The flask or serum bottle procedure, on the other hand, is a widely applicable method for assessing biodegradability. The test is simple, equipment needs are minimal, and many versions of this test are in use. The Soap and Detergent Association has standardized the test for use under aerobic conditions,[20] while Owens *et al.* applied the test for use under anaerobic conditions, and referred to it as the biochemical methane potential test.[21] This test has been further modified and used successfully for comparing the biodegradation potential of several biomass feedstocks, evaluating microbial inocula and pretreatment techniques, developing biodegradability constants, and determining factors contributing to substrate inhibition.[7] In these flask tests, sample biodegradability is determined by measuring the cumulative methane production and organic matter degraded from an incubated sample. The results can be expressed as the amount of methane produced per weight of organic matter, e.g. standard cubic meters per kilogram (SCF/lb) VS added, and/or the percent of organic matter destroyed (e.g. VS reduction). Relative rates of biodegradability can also be compared using this assay. Figure 5.1, for example, shows the application of the assay to determine methane production rates and methane yields from two substrates, sewage sludge and water hyacinth.

The conditions of this biodegradability assay should be established to promote reproducible results using the same substrate and among different assays. Physical and chemical parameters must be carefully controlled to ensure the growth and metabolic activity of a healthy microorganism population. The optimum substrate and inoculum size should also be established to ensure a balanced fermentation and to reduce the potential for toxicity.

Mathematical models have been applied to substrate composition analyses discussed earlier to evaluate the nutritive value and to quantify

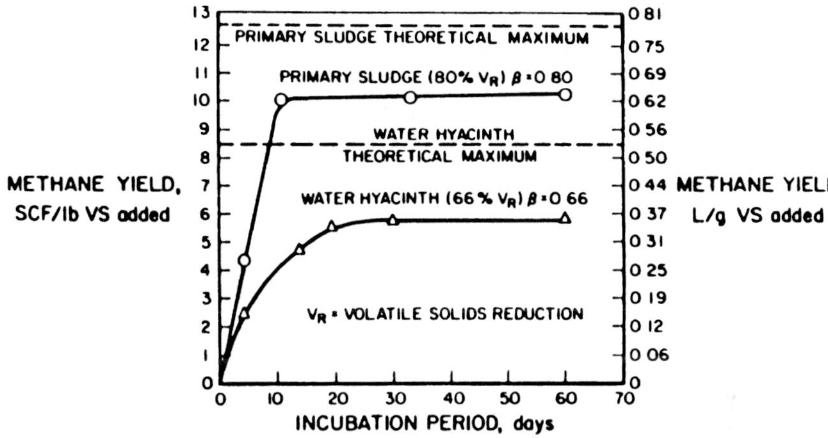

FIG. 5.1. Anaerobic biogasification potential of water hyacinth and primary sludge (ABP assay 35 °C).

the digestibility of animal feeds.[22] Chandler, for example, employed these analyses and models to predict methane fermentation biodegradability.[8] By using the physical–chemical analyses of a large number of diverse substrates, an accurate prediction of substrate biodegradability (eqn (5.2)) could be made. These predictions can be useful when long-term batch biodegradation studies are prohibitive.

$$B = 0{\cdot}028x + 0{\cdot}830 \tag{5.2}$$

where B = biodegradable fraction, and x = VS lignin content (%).

5.5 EFFECT OF COMPOSITION ON BIODEGRADABILITY

5.5.1 Introduction

The process of biodegradation is the result of a highly complex chain of events revolving around interactions between the specific compound, microorganisms, and the environmental conditions where the two are found.[23] Biodegradation is defined as a change in the physical or chemical properties brought about by biological activity, accompanied by a decrease in the energy of the parent compound.[24] The degree of change can be classified as primary biodegradation, which is the transformation of the parent compound so as to be no longer detectable as the parent compound (that is, cellulose to glucose), or ultimate biodegradation, which is the conversion of the parent compound to CH_4 and CO_2 under anaerobic conditions (for example, cellulose to methane).

A compound that is not biodegradable or persists for long periods is termed recalcitrant or refractory. Certain natural organic compounds, such as lignin, are refractory to decomposition under anaerobic conditions even at long residence times. This explains why organic matter accumulates as lake sediment and fossil fuels such as peat, oil, and coal. Generally, this refractory property is related to the lack of enzymes in the anaerobic bacteria or an oxygen requirement by many enzymes to carry out the initial hydrolytic reactions.[25] It has been established that certain compounds can complex with other compounds in such a manner as to render them refractory to anaerobic decomposition. For example, lignin affects the decomposition of cellulose in this manner.

Microbial acclimation and co-metabolism are important factors in determining the biodegradability of various compounds. Acclimation of bacteria to metabolize a specific substrate occurs by mutation and natural selection or by induction of specific enzymes. Many synthetic compounds, such as PCB and DDT, have been successfully degraded through bacterial acclimation; however, increased biodegradability of biomass feeds via microbial acclimation has not been reported. IGT has evaluated several marine inocula for the digestion of sea kelp and has not observed greater degradation than that observed with a sludge inoculum.[26] Co-metabolism involves the degradation/transformation of a secondary compound while another compound is being degraded as the primary carbon and energy source. This process has been studied for wastewater treatment but has received little attention in biomass digestion.[27] Substrate biodegradability can have a considerable impact on the economics of methane fermentation because higher methane yields (biodegradability) can significantly reduce the high costs of biomass production.[28]

5.5.2 Chemical Composition

Lignocellulosics contain cellulose, hemicellulose, and lignin as three major components. Cellulose is a linear homopolymer of anhydroglucose units linked by β-D-1,4-glucosidic bonds. Without side chains, cellulose molecules are sterically compatible for close association and in forming intra- as well as inter-molecular hydrogen bonding. Cellulose is generally crystalline with up to about 15% being 'amorphous'. The exact crystalline structure of cellulose is still a subject of debate among carbohydrate scientists. Pertinent to biological methane generation is the resistance, due to the structure, toward hydrolysis and other chemical conversion. Pure starch is digestible by numerous living organisms. Pure cellulose is much more resistant, not necessarily due to the β-glucosidic bonds being different from those of the α configuration, but rather to the crystalline structure, which makes cellulose difficult to hydrolyze not only in a thermodynamic sense but also in the sense of reaction kinetics. The heat of cellulose crystallization has been the subject of a recent doctoral thesis.[29] The subject of surface accessibility of cellulose to binding by a cellulolytic enzyme is the topic of another doctoral thesis.[30] The effects of several structural parameters on cellulose hydrolysis, including surface, pore size, moisture content, and degree of polymerization, are reviewed by numerous authors.[31–34]

Hemicellulose is associated with cellulose and lignin and supposedly furnishes cell wall rigidity and flexibility. Hemicellulose is not a homopolymer but rather is complex in terms of both monomeric components and molecular structures. Generally speaking, hemicelluloses are a family of highly branched polymers. In hydrolyzates of hemicellulose, a number of pentoses, hexoses, and uronic acids have been identified including D-xylose, D-mannose, D-glucose, D-galactose, L-arabinose, D-glucuronic acid, 4-O-methyl-D-glucuronic acid and D-galacturonic acid.[35] Based upon their constituents, hemicelluloses have been classified into two groups, pentosans and hexosans, which exist in variable proportions of hemicellulose from native plant materials without much depolymerization. Many microbes can utilize the hemicellulose fraction in either the native polymeric form or after a prior hydrolysis. Thus, hemicellulose causes no serious resistance to conversion of cellulosics into chemicals, including methane.

Even though up to about 90% of the available hemicellulose can be readily extracted and/or hydrolyzed, there could be 10% or so of the total hemicellulose in the form of a linear polymer, xylan, which is closely associated with linear cellulose molecules and is resistant to hydrolysis. In an integrated process of utilization of cellulosics, Tsao *et al.*[36] described a prehydrolysis treatment to extract and remove hemicellulose by a mild acid. It is followed by other pretreatments to render cellulose more hydrolyzable. One should expect a measurable amount of D-xylose in the cellulose hydrolyzates even after the prehydrolysis of hemicellulose due to this resistant residual xylan chain.

Lignin is a polyphenolic macromolecule and its chemical structure is not yet completely clear. The basic units of plant lignin are thought to be

3,5-dimethoxy-4-hydroxyphenylpropane, 3-methoxy-4-hydroxyphenyl-propane, and 4-hydroxyphenylpropane. The chemistry and physiology of degradation of lignin by living cells should be of great interest to biological methane generation. First of all, lignin is rich in carbon and hydrogen and relatively poor in oxygen, and thus could be a desirable raw material for CH_4 generation. However, lignin is thought to be refractory under anaerobic conditions.[37] Meanwhile, biological degradation of lignin by microorganisms as a possible pretreatment of cellulosic materials for subsequent cellulose and hemicellulose utilization has been proposed.[38,39] We therefore have reasons for a better understanding of the biochemistry involved in lignin degradation. In recent years, some of the fractions comprising degradation have been elucidated, and the unusual biochemical and physiological features are beginning to be described. Recent literature on this subject is quite comprehensive, including several reviews.[39–46]

Lignin removal from native cellulosics by solvent extraction has also been proposed as a potential pretreatment for cellulose utilization. Indeed, if lignin contributes little as a carbon and hydrogen source for methane generation and yet efficient methods are available for its separation from cellulose and hemicellulose, the need for an integrated scheme including utilization of lignin coproducts will arise for reasons of economics and process efficiency, as well as for the alleviation of a waste disposal requirement that would otherwise exist.

5.5.3 Physical Structure

The crystalline structure of cellulose molecules is not yet completely clear, as mentioned above. In addition, several levels of architectural structures also exist in cellulosic materials and are pertinent to their conversion. An outline of a group of plant cells is shown in Fig. 5.2. The cellulosics are the cell wall materials of plants. Surrounding the cell is the middle lamella, which is heavily lignified and contains lignin and hemicellulose at approximately 70% and 30%, respectively. Cellulose is contained in the cell wall. Within the cell wall, four distinct concentric morphological layers can be distinguished. The outermost layer is called the primary wall, which is about 0·1 mm thick and contains about one-third cellulose. In the primary wall, the microfibrillar structure appears as a loose and random network. The microfibrils are oriented around the fiber at approximately right-angles to the axial fiber direction. The secondary wall forms during growth and maturation of the cell and contains three layers: an outer layer, S1; a middle layer, S2; and an inner

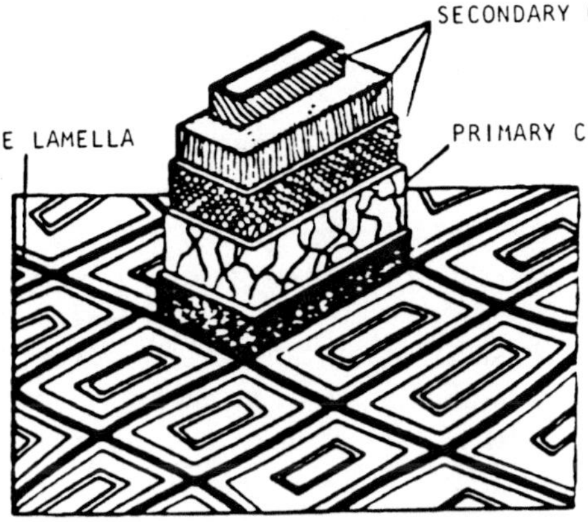

FIG. 5.2. Architecture of plant cell wall.

layer, S3. The microfibrils of the outer layers are wound in flat helices. The direction of winding alternates to form overlapping spirals. In the middle layer, S2, the cellulose fibrils are tightly packed in a steep parallel helix. The innermost, S3, layer consists of helical microfibrils. The structure of this layer is not clear. For a raw cotton fiber, the primary wall is about 0·1–0·2 μm thick; the secondary wall is 1–4 μm thick and contains almost pure cellulose. In general, cellulose occupies about 90% of the dry weight of the secondary wall in various fiber cells. The primary element of the supermolecular structure of cellulose is elementary microfibrils which are bound by linkage of individual linear polymeric cellulose molecules. X-ray diffraction and other evidence indicate that cellulose microfibrils are partially crystalline and partially disordered.

The above-mentioned structural features are pertinent to biological conversion of cellulosics in several ways. In enzymatic hydrolysis, the prerequisite for a reaction to take place is a direct association between the molecules of the enzyme and the substrates. In this case, the barrier represented by the lignin-rich middle-lamella makes the diffusion and the penetration by the macromolecular enzymes extremely difficult and slow, if not impossible. The phenolic groups in lignin might even be inhibitory toward enzymes. The enhancement in biodigestibility of cellulosics by a simple pretreatment of size reduction might be explained in terms of the more accessible ends of the long cellulose microfibrils wrapped inside lignin barriers, as well as the rupture of the barrier by the cutting action. The lignin layer is expected to cause less of a barrier problem for acid hydrolysis because of the relatively small molecular size of the acid catalysts. This size difference between the acid catalysts and the enzymes might help explain the difference in effectiveness of the same pretreatment for cellulose hydrolysis when different catalysts are employed.

When anaerobic microbes are attacking cellulosics directly, they may excrete enzymes as extracellular agents to create soluble, metabolizable glucose and other carbohydrates that can then be utilized by the cells after uptake. However, direct attachment of microbial cells onto the surface of the solid substrate should enhance the rate of the degradation process solely on the basis of mass transfer considerations if not other, perhaps deep-rooted but unelucidated, physiological reasons. A recent review by Moo-Young *et al.*[47] provides a comprehensive discussion of solid-substrate fermentations.

5.5.4 Biodegradability of Different Feedstocks

The biodegradability of a broad spectrum of biomass feedstocks expressed as methane yield and reduction in organic matter is presented in Table 5.5. These data approach ultimate yields since they were obtained from reactors operated at long retention times and low loading rates. Most of the non-woody feeds were about 60–80% biodegradable, with methane yields ranging from 0·25 to 0·35 SCM/kg (4·0 to 5·6 SCF/lb) VS added. Some sorghum hybrids, however, were more than 90% biodegradable, the highest reported for herbaceous species without pretreatment.[18] Several recent studies have demonstrated significant anaerobic conversion (up to 60%) of different hardwood species (alder, cottonwood, poplar, and sycamore) without pretreatment.[51,52] These studies indicate that the biogasification of wood is a technically feasible process.

Herbaceous biomass is composed of approximately 60–70% lignocellulosic material compared to about 95–99% for woody biomass, with soluble organic matter accounting for the remaining material in both groups. Although pure cellulose is biodegradable under anaerobic conditions,[51] the hydrolysis of cellulose, either purified or in a lignocellulosic matrix, has been cited as the rate-limiting step in its conversion to methane. Accumulation of cellulose hydrolysis products, cellobiose and glucose, has not been observed as these compounds are rapidly fermented.[53] The rate of cellulose decomposition is reportedly influenced by such characteristics as the degree of crystallinity, capillary structure, and degree of polymerization of the molecule.[54] Hemicellulose, a highly branched heteropolymer, is readily utilized by microorganisms and yields a variety of hexoses and pentoses upon hydrolysis. These structural carbohydrates, however, become increasingly resistant to microbial

TABLE 5.5
Biodegradability of Biomass Feedstocks*

Feedstock	*Experimental methane yield, SCM/kg VS added*	*Experimental methane yield, SCF/lb VS added*	*Volatile solids reduction, %*
Aquatic			
Sea kelp	0·20	3·2	48
Sea kelp	0·37	5·9	—
Sea kelp	0·27	4·3	62
Water hyacinth	0·34	5·4	66
Water hyacinth	0·24	3·8	59
Laminaria saccharina	0·30	4·8	58
Herbaceous—energy crops			
Bamboo	0·17	2·7	41
Bermuda grass	0·32	5·1	69
Cattail	0·27	4·3	51
Cattail	0·22	3·5	59
Giant reed	0·33	5·3	72
Kudzu	0·29	4·6	64
Napier grass	0·32	5·1	74
Sorghum (Johnson grass)	0·38	6·1	79
Sorghum (RIO)	0·40	6·4	92
Sorghum (623 X RIO)	0·34	5·4	91
Herbaceous—agricultural residues			
Corn stover	0·22	3·5	50
Corn stover	—		71
Corn stover	0·27	4·3	62
Corn stover	0·18	2·9	41
Corn stalks	0·29	4·6	77
Corn leaves	0·27	4·3	72
Wheat straw	0·22	3·5	55
Wheat straw	0·26 (55 °C)	4·2	—
Woody			
Alder, black	0·24	3·8	33
Alder, red	0·28	4·5	48
Cottonwood	0·22	3·5	33
Eucalyptus	0·014	0·22	21
Hybrid poplar	0·32	5·1	54
Hybrid poplar	—	—	10
Loblolly pine	0·063	1·0	3·6
Pine	—	—	5·8
Sycamore	0·32	5·1	57
White fir	—	—	5·0

*Data from Refs 8, 16, 17, 18, 48, 49, and 50.

attack when associated with lignin in a lignocellulosic matrix.[55,56] A linear relationship, for example, was demonstrated between VS reduction efficiency and lignin content to predict methane fermentation biodegradability.[8] The amount of lignocellulosic materials has been shown to govern the rate of hydrolysis more than either reduction in particle size or reduction in crystallinity.[57] Although lignin is reportedly resistant to microbial degradation in anaerobic environments,[25] the aromatic monomers comprising the lignin nucleus have been degraded anaerobically.[58] Although previous data suggest that woody biomass has a low methane fermentation potential due to high lignin concentrations, more recent data suggest that the association between lignin and other cell wall polymers may not be as detrimental to microbial enzyme attack as previously reported.[51,52] The anaerobic conversion efficiencies achieved from woody substrates at long solids retention times and with added macro- and micronutrients indicated that approximately 85% of the structural carbohydrates were converted to methane, assuming that lignin is refractory. Studies comparing methane yields from herbaceous and woody biomass at long solids retention times did not demonstrate a linear relationship between lignin concentration and anaerobic biodegradability. Although lignin concentrations may have an effect on the rate of methane fermentation,[52] other cellular components may be more directly related to the methane fermentation potential of biomass.

The cellular material not associated with the cell wall matrix consists of water-soluble and nonstructural carbohydrates as well as nitrogenous compounds such as proteins and nucleic acids. Sucrose is the most abundant storage sugar found in herbaceous plants, with concentrations ranging from 40% to 60% for sweet sorghum and energy cane.[59] Such high sucrose concentrations contribute to the high biodegradability of those forms of biomass. The mannitol content of kelp was shown to have a significant impact on the fraction of the theoretical methane yield that was achieved experimentally. The methane yields achieved from kelp lots with mannitol contents ranging from 5% to 30% were shown to increase logarithmically as the mannitol content increased (Fig. 5.3).[60] An equation to predict this relationship has been derived and verified experimentally.

High-correlation regression equations were derived relating concentration of soluble carbohydrates and protein to methane yields for several marine and terrestrial biomass species.[61] The soluble carbohydrates in the samples tested constituted less than 25% of the dry weight of the feedstocks, while the protein concentrations were lower.

Of the nitrogenous compounds in plants, 60–80% constitutes true

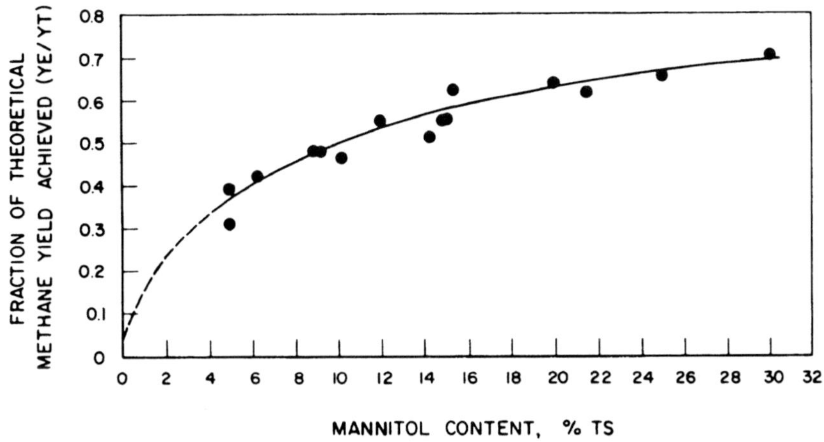

FIG. 5.3. Fraction of theoretical methane yield achieved experimentally at different kelp mannitol concentrations (semicontinuous—Fed STR, 0·1 lb VS/ft^3-day loading, 15–18 day SRT, and 35 °C) (data from Ref. 60).

protein, with the remainder being water soluble and insoluble nonprotein nitrogen.[21] Crude protein concentrations in herbaceous biomass have been estimated to range from 5% to 30% of the VS fraction. Following 120 days' batch fermentation, approximately 60% of the crude protein was reported to be fermentable, indicating that a large percentage of the crude protein was refractory.[8]

The relative rates of feed biodegradation are also affected by feed composition. The biogasification potential of different biomass and waste feeds was examined under identical conditions using the anaerobic biogasification potential assay. Figure 5.4 shows that the slopes of the curves vary, indicating a difference in the relative rates of biodegradability. These differences were confirmed by determining certain kinetic parameters, maximum specific growth rate and substrate saturation coefficient, during the operation of a CSTR with water hyacinth, primary sludge, and kelp (Table 5.6).

The digester gas composition is reportedly a function of the oxidation state of the carbon in the biomass, the CO_2 saturation of the digester lignin, and the nitrogen content of the biomass (Fig. 5.5).[62] Digester gas produced from feeds containing high fat concentrations, for example, is expected to have a higher methane content than gas produced from urea-containing feeds. The observation partially explains the high percent

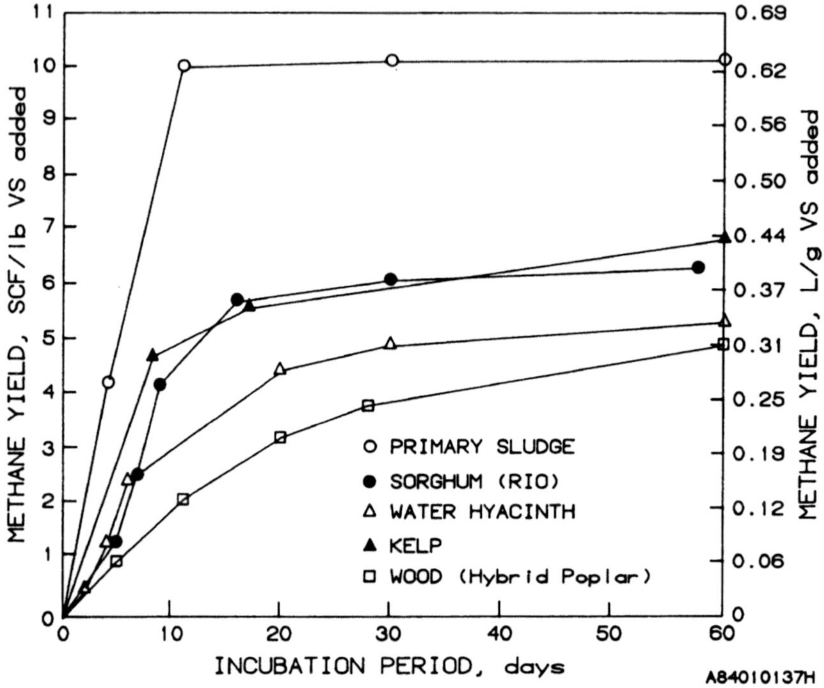

FIG. 5.4. Biogasification of biomass feedstocks (ABP assay 35 °C).

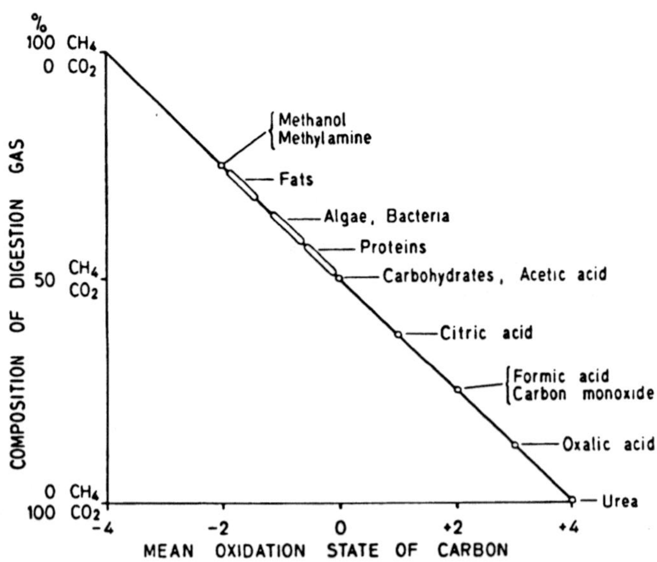

FIG. 5.5. Composition of the digestion gas depending on the mean oxidation state of the carbon in the substrate, assuming total mineralization of the substrate.

TABLE 5.6
Kinetic Parameters for the Digestion of Water Hyacinth, Primary Sludge and Kelp in a CSTR (35 °C)

Feed	*Maximum specific growth rate* (μ), *days*$^{-1}$	*Saturation coefficient* (k), *g/liter*
Water hyacinth	0·25	11·0
Primary sludge	1·32	23·1
Kelp	1·72	129·3

of methane frequently observed during the digestion of high-fat-containing sludges.

5.6 SUMMARY

The major factor affecting the suitability and performance of biomass in anaerobic digestion is its composition. The solids content of the biomass affects digester design, methods of digester operation, and materials-handling procedures. Theoretical methane yields can be predicted through elemental analyses or through determination of the biomass heating value. Accurate prediction of the expected methane yield from biomass requires a determination of the biodegradable organic matter as well as the fraction of the biomass that will be consumed for synthesis and maintenance of the microbial population.

Of the several methods that are available to determine the organic content of biomass (for example, volatile solids, COD, TOC, and BOD) only those methods that involve biological activity can adequately determine its biodegradability. Although the BOD test measures the potential for substrate oxidation by biological processes, a more appropriate test, the anaerobic biogasification potential assay, has been developed for measuring the anaerobic biodegradability of biomass using batch-fed culture techniques. The test is simple and equipment needs are minimal. Culture conditions provide long incubation periods and an excess of nutrients and growth factors. The batch bioassay should be standardized for determining ultimate biodegradability of feedstocks suitable for anaerobic digestion.

During anaerobic digestion, the amount of biomass substrate needed to maintain the population of digesting microorganisms depends upon the nature of the biomass and upon such operating conditions as SRT and MRT.

The biomass feed fraction converted to bacterial cells is readily determined for soluble feeds. However, a reliable technique is needed to estimate the cell maintenance fraction for the digestion of particulate feeds in the several reactor types that may be used in biomass digestion. Accurate determination of this parameter will help to enable more rational predictions of methane yields and determine the reactor designs and operating conditions needed for maximum methane yields.

The methane potential and biodegradability of biomass varies substantially among different species, and also within the same species when growth conditions vary. Non-lignin-containing plants, such as kelp, are generally more biodegradable than lignin-containing plants, such as hybrid poplar.

Further work is needed to determine the optimal growth conditions required for producing biomass that is highly suitable for anaerobic digestion processes. This work should be directed toward improving the methane yields, the rate of biodegradability, and process stability.

Biomass represents a varied and complex source of organic matter. Many species possess recoverable components with good economic value. To improve the efficiency of biomass digestion processes and to reduce potential waste disposal problems, digester residues should be utilized for their economic value. Further work is needed to determine the economic benefits of coproduct utilization through the implementation of integrated biomass production, processing, and recovery schemes.

REFERENCES

1. *Standard Methods for the Examination of Water and Wastewater.* Washington, DC: American Public Health Association, 15th edn, 1982.
2. *1978 Annual Book of ASTM Standards, Part 26: Gaseous Fuels; Coal and Coke; Atmospheric Analysis.* Philadelphia: American Society for Testing and Materials, 1978.
3. *Official Methods of Analysis of the Association of Official Agricultural Chemists*, 12th edn, Washington, DC: Association of Official Agricultural Chemists, 1975.
4. Chynoweth, D. P., 'Microbial Conversion of Biomass to Methane.' Paper presented at the *Eighth Annual Energy Technology Conference and Exposition*, Washington, DC, March 1981.
5. Symons, G. E. and Busewell, A. M., 'The Methane Fermentation of Carbohydrates,' *J. Am. Soc.* **55**:2028–2029, 1933.

6. Speece, R. E. and McCarty, P. C., 'Nutrient Requirements and Biological Solids Accumulation in Anaerobic Digestion,' in *Adv. in Wat. Poll. Res.*, Proceedings of the First International Conf., Vol. **2**:305–322, 1964.
7. Chynoweth, D. P., Jerger, D. E., Conrad, J. R., Razik, A., Srivastava, V. J., Ghosh, S., Henry, M. P. and Babu, S., 'Gasification of Land-Based Biomass.' Final Project Report for GRI, IGT Project No. 30564, Chicago, 1983.
8. Chandler, J. A., 'Predicting Methane Fermentation Biodegradability,' MS Thesis, Cornell University, 1980.
9. IGT, unpublished data.
10. Van Soest, P. J., 'Use of Detergents in the Analysis of Fibrous Feeds, Parts I and II,' *Assoc. of Agr. Chem. J.* **46**:825–835, 1963.
11. Hanson, R. S. and Philips, J. A., 'Chemical Composition,' Chapter 17 in Section IV — Metabolism, in W. A. Wood, Ed., *Manual of Methods for General Bacteriology*. Washington, DC: American Society for Microbiology, 1981.
12. McNair, H. M. and Bonelli, E. J., *Basic Gas Chromatography*. Palo Alto, CA: Varian Associates, 1969.
13. Johnson, E. L. and Stevenson, R., *Basic Liquid Chromatography*. Palo Alto, CA: Varian Associates, 1978.
14. Ghosh, S. and Henry, M., 'Application of Packed-Bed Upflow Towers in Two-Phase Anaerobic Digestion,' in Y. C. Wu *et al.*, Eds, *Proc. 1st Inter. Conf. on Fixed-Film Biolog. Proc.*, Kings Island, Ohio, 1982.
15. VanVelson, A. F. M. and Lettinga, G., 'Effect of Feed Composition on Digester Performance,' pp.113–130 in *Proceedings of First International Symp. on Anaerobic Digestion*. London: Elsevier Applied Science Publishers, 1980.
16. Van Soest, P. J., Winr, R. H. and Moore, L. A., 'Estimation of the True Digestibility of Forages by the *In Vitro* Digestion of Cell Walls,' pp.438–441 in *Proceedings of the 10th Intl. Grassland Congr.*, 1966.
17. Mertens, D. R., 'Application of Theoretical Mathematical Models to Cell Wall Digestion and Forage Intake in Ruminants.' PhD Thesis, Cornell University, 1973.
18. Millett, M. A., Baker, A. J. and Satter, L. D., 'Pretreatments to Enhance Chemical Enzymatic and Microbiological Attack of Cellulosic Materials,' in *Biotechnol. and Bioeng. Symp. 5: Cellulose as a Chemical and Energy Resource*. New York: John Wiley & Sons, 1975.
19. Barth, E. F. and Bunch, R. L., 'Biodegradation and Treatability of Specific Pollutants.' *EPA-600/79-034*, 1979.
20. Soap and Detergent Association, 'A Procedure and Standards for the Determination of the Biodegradability of Alkyl Benzene Sulfonate and Linear Alkylate Sulfonate.' *J. Am. O.L. Chem.* **42**, November 1965.
21. Owens, W. F., Stuckey, D. C., Healey, J. B., Young, L. Y. and McCarty, P. L., 'Bioassay for Monitoring Biochemical Methane Potential and Anaerobic Toxicity,' *Wat. Res.* **13**:485–492, 1979.
22. Van Soest, P. J., 'Development of a Comprehensive System of Feed Analysis and Its Application to Forages,' *J. Am. Sci.* **269**, **1**:119–128, 1967.
23. Kirsch, E. J., Grady, C. P. L. and Wukasch, R. F., 'Protocol Development for the Prediction of the Fate of Organic Priority Pollutants in Biological Wastewater Treatment Systems,' US EPA, March 1981.

24. Standard Methods Committee on Biodegradability, Required Characteristics and Measurement of Biodegradability, *JWPCF* **39**:7, 1232–1235, 1967.
25. Zeikus, J. G., 'Fate of Lignin and Related Aromatic Substances in Anaerobic Environments,' in *Lignin Biodegradation: Microbiology, Chemistry and Potential Applications*, Vol. I. Boca Raton, Fla.: CRC Press, 1980.
26. Chynoweth, D. P., 'Biomethanation of Giant Brown Kelp, *Macrocystis pyrifera*,' in Symposium Papers, *Energy From Biomass and Wastes III*. Lake Buena Vista, Fla.: Institute of Gas Technology, 1978.
27. Lavalle, K., 'Biodegradation of Organic Compounds by Anaerobic Processes', in A. Obayaski and J. Gorgan, Eds, *Management of Industrial Pollutants by Anaerobic Processes*. Chicago: IIT, 1982.
28. IGT, unpublished data.
29. Dale, B. E., 'A Microcalorimetric Study of Cellulose and Its Metal Complex Solvents,' PhD Thesis, Purdue University, 1979.
30. Gilbert, I. G., 'A Study of Enzyme Adsorption and Reaction Kinetics for Cellulose Hydrolysis,' PhD Thesis, Purdue University, 1982.
31. Cowling, E. G. and Brown, W., 'Structural Features of Cellulosic Materials in Relation to Enzymic Hydrolyis,' *Adv. in Chem. Ser.* **95**:152–187, 1969.
32. Cowling, E. G. and Kirk, T. K., 'Properties of Cellulose and Lignocellulosic Materials as Substrates for Enzymatic Conversion Processes,' *Biotech. Bioeng. Symp.* **6**:95–123, 1976.
33. Fan, L. T., Lee, Y. H. and Berdmore, D. H., 'Major Chemical and Physical Features of Cellulosic Materials as Substrate for Enzymatic Hydrolysis,' *Adv. in Biochem. Eng.* **14**:101–117, 1980.
34. Chang, M. M., Chou, T. Y. C. and Tsao, G. T., 'Structure, Pretreatment, and Hydrolysis of Cellulose,' *Adv. Biochem. Engr.* **20**:15–42, 1981.
35. Whistler, R. L. and Richards, E. L., Hemicellulose, *The Carbohydrates, IIA*, p. 447 Pigman and Horton, Eds. New York: Academic Press, 1970.
36. Tsao, G. T., Ladisch, M., Ladisch, C., Hsu, T. A., Dale, B. and Chou, T., 'Fermentation Substrates From Cellulosic Materials: Production of Fermentable Sugars From Cellulosic Materials,' *Annual Reports Ferment. Processes* **2**:1–22, 1978.
37. Zeikus, G., 'Fate of Lignin and Aromatic Substances in Anaerobic Environments,' pp. 101–110 in *Lignin Biodegradation: Microbiology, Chemistry Potential Applications*, Vol. II. Boca Raton, Fla: CRC Press, 1980.
38. Brooks, R. E., Bellamy, W. and Su, T., 'Bioconversion of Plant Biomass to Ethanol,' *NTIS*, COO4147/4, 1978.
39. Ander, P. and Eriksson, K. E., 'Lignin Degradation and Utilization by Microorganisms,' *Progr. Indust. Microbiol.* **14**:1–58, 1978.
40. Amer, G. I. and Drew, S. W., 'Microbiology of Lignin Degradation,' *Annual Reports Ferment. Processes* **4**:67–103, 1980.
41. Crawford, D. L. and Crawford, R. L., 'Microbial Degradation of Lignin,' *Enzyme and Microbial Technol.* **2**:439–444, 1980.
42. Kirk, T. K., 'Toward Elucidating the Mechanism of Action of the Ligninolytic Systems in Basidiomycetes,' pp. 131–149 in Hollaender, Rabson, Roger, San Pietro, Valentine and Wolfe, Eds, *Trends in the Biology of Fermentations*. New York: Plenum Press, 1981.

43. Kirk, T. K., 'Degradation of Lignin,' in Gibson, Ed., *Biochemistry of Microbiological Degradation.* New York: Marcel Dekker, 1982.
44. Kirk, T. K., 'Degradation and Conversion of Lignocellulose.' *The Filamentous Fungi* **4**, 1983.
45. Kirk, T. K., Higuchi, T. and Chang, H. M., *Lignin Biodegradation: Microbiology, Chemistry and Potential Applications*, Vols I and II. Boca Raton, Fla: CRC Press, 1980.
46. Crawford, R. L., *Lignin Biodegradation and Transformation.* New York: Wiley Press, 1981.
47. Moo-Young, M., Moreira, A. R. and Tengerdy, R. P., 'Principles of Solid-Substrate Fermentation.' *The Filamentous Fungi* **4**:115–144, 1983.
48. Fannin, K. F., Srivastava, V. J., Mensinger, J. D. and Chynoweth, D. P., 'Marine Biomass Program: Digestion Systems Development and Stability Study.' IGT Project No. 30558 for Gas Research Institute, Contract No. 5082-225-0607, Chicago, 1983.
49. Gaddy, J. L., Foutch, G. L. and Book, N. L., 'Potential of Corn Stover as a Fuel.' Paper presented at AIChE Meeting, San Francisco, 1979.
50. Pfeffer, J. T., 'Biological Conversion of Biomass to Methane, Corn Stover Studies,' Project Report for December 1977–August 1978. US DOE Contract No. EY-76-S-02-2197, University of Illinois, 1979.
51. Jerger, D. E., Conrad, J. R., Fannin, K. F. and Chynoweth, D. P., 'Biogasification of Woody Biomass,' pp. 341–372 in Symposium Papers, *Energy From Biomass and Wastes VI*, Chicago: Institute of Gas Technology, IL, 1982.
52. Jerger, D. E., Dolenc, D. A. and Chynoweth, D. P., 'Bioconversion of Woody Biomass as a Renewable Source of Energy,' *Biotechnol. Bioeng. Symp. No. 12*, 233–248, 1982.
53. Bull, M. J., *Progress in Industrial Microbiology*, Vol. 16. New York: Elsevier Scientific Publishing Co., 1982.
54. Cowling, E. B., 'Physical and Chemical Constraints in the Hydrolysis of Cellulose and Lignocellulosic Materials,' *Biotechnol. Bioeng. Symp.* **5**:163, 1975.
55. Baker, A. J., Millett, M. A. and Satter, L. D., 'Wood and Wood-Based Residues in Animal Feeds,' in ACS Symposium Series 10: *Cellulose Technology Research.* Washington, DC: American Chemical Society, 1975.
56. Scheffer, T. C., 'Natural Resistance of Wood to Microbial Deterioration,' *Annual Review of Phytopathology* **4**:147–170, 1966.
57. Fan, L. T., Lee, Y. and Beardmore, D. H., 'Mechanism of the Enzymatic Hydrolysis of Cellulose: Effects of Major Structural Features of Cellulose on Enzymatic Hydrolysis,' *Biotechnol. and Bioeng.* **22**:177–199, 1981.
58. Healy, J. B., Jr and Young, L. Y., 'Anaerobic Biodegradation of Eleven Aromatic Compounds to Methane,' *Appl. and Environ. Microbiol.* **38**:1, 84–89, 1979.
59. Elawad, S. H., Gascho, G. J. and Shih, S. F., 'The Energy Potential of Sugarcane and Sweet Sorghum,' in Symposium Papers, *Energy From Biomass and Wastes VI.* Chicago: Institute of Gas Technology, April 1981.
60. Fannin, K. F., Srivastava, V. J. and Chynoweth, D. P., 'Unconventional

Anaerobic Digester Designs for Improving Methane Yields From Sea Kelp,' pp. 373–399 in Symposium Papers, *Energy From Biomass and Wastes VI*. Chicago: Institute of Gas Technology, 1981.
61. Habig, C. and Rhyther, J. H. 'Some Correlations Between Substrate Compositions and Biogas Yields,' pp. 817–832 in Symposium Papers, *Energy From Biomass and Wastes VIII*. Chicago: Institute of Gas Technology, 1984.
62. Gujer, W. and Zehnder, A. J. B., 'Conversion Processes in Anaerobic Digestion,' *Water Science Technology*, **15**:127–167, 1983.

6
Pre-/Posttreatment

GEORGE T. TSAO

School of Chemical Engineering, Purdue University, Lafayette, Indiana, USA

6.1 INTRODUCTION

A significant fraction of many biomass energy feedstocks may be refractory to anaerobic decomposition. Pretreatment by physical, chemical and biological processes may be of key importance in unlocking the huge amounts of cellulosic renewable resources to assure a high process efficiency in anaerobic digestion. Certain biomass feed sources such as cheese whey and starchy wastes can be readily fermented by microorganisms to generate various products including methane gas. Pretreatment is not necessary with such sources because starch, lactose, and soluble proteins can be readily metabolized by many living organisms.

This chapter will describe various techniques of pretreatment that have been attempted to overcome the difficulties in biodigestion caused by the natural characteristics of cellulosic materials. The majority of the pretreatment literature has been directed toward the saccharification of cellulosics necessary to provide a suitable substrate for alcohol production. Complete saccharification of the feeds may not be necessary, however, for methane production. Several new pretreatment methods for pretreatment of cellulosic material will also be discussed.

6.2 PRETREATMENT TECHNIQUES

Cellulosics are generally resistant to hydrolysis by enzymes or acids because of two major hindrances related to cellulose structure and the lignin barrier. Other factors such as accessible surface, pore size distribution, degree of polymerization and moisture content also affect the rate and the extent of cellulose digestion. Pretreatment methods are employed to modify these physical and chemical characteristics in the hope of enhancing the conversion of cellulosics, particularly cellulose hydrolysis. The effectiveness of a pretreatment should be evaluated in more ways than one. It may enhance the rate of cellulose hydrolysis and thus also the rate of methane generation. It may also affect the extent of cellulose hydrolysis and thus the ultimate yield of the final products. With acquisition of a fundamental understanding of the physical, chemical, and structural nature of the cellulosics, it is possible to evaluate the effectiveness of a pretreatment method according to selected criteria of rate, yield, and other measurements. Furthermore, a good understanding of the fundamentals of pretreatment processes will help in designing better pretreatment methods.

Several recent reviews dealing with various pretreatment techniques have been authored by Millet *et al.*;[1] Dunlop *et al.*;[2] Tsao and Chiang;[3] and Jerger *et al.*[4] Some selected pretreatment methods are summarized in Table 6.1, and discussed below.

6.3 SIZE REDUCTION

Cellulosic materials have been described as fibrous cellulose micofibrils wrapped in a continuous barrier of middle lamella consisting of lignin and hemicellulose. In the native state, the accessibility to cellulose is presumably limited to only the openings at both ends of a long and thin structure. Size reduction is viewed to increase the number of exposed ends by cutting the long structure into many short ones. If this conceptual view is correct, size reduction should be expected to enhance the rate more than the yield of hydrolysis. Indeed, this has been reported by Jerger *et al.*,[4] who used milling and wet extrusion milling to reduce the size of cellulosics to 0·003–8 mm and found that the methane generation rate was significantly affected by the particle size (Fig. 6.1).

However, much depends upon the techniques of milling, some of which may contribute more than just simple size reduction. Ball milling, for example, has been reported to decrease the size and also the crystallinity of cellulose; thus, this treatment can enhance not only the rate, but also the ultimate glucose yield of cellulose hydrolysis. Pew and Weyna[5] obtained essentially complete cellulose digestion of spruce and aspen after ball milling compared with less than 10% digestion of the unpretreated material. Tassinari and Macy[6] applied differential speed,

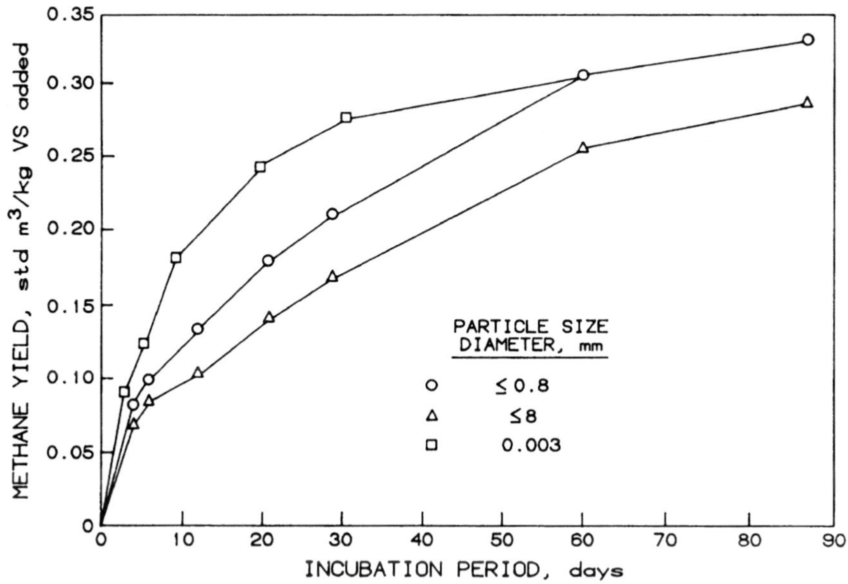

FIG. 6.1. Methane yields from hybrid poplar at various particle sizes (batch digestion, 35 °C).[4]

two-roll milling and achieved an impressive increase in cellulose hydrolysis by enzymes. Nystrom[7] also showed an almost linear increase in enzymatic susceptibility of cellulose with milling time.

Recently, Kelsey and Shafizahed[8] applied simultaneous wet milling of various cellulosic materials during enzymatic hydrolysis and observed an extraordinary increase in the rate of glucose formation by increased accessibility; the mechanical force exerted on the cellulose substrate during simultaneous milling and hydrolysis may enhance the hydrolysis by two additional mechanisms, as speculated by Tsao and Chiang.[3] The force may 'bend' or distort the structure such that macromolecular enzymes can readily penetrate into the interior of the crystalline cellulose microfibrils. The force may also help peel off the surface layers of oligomeric anhydroglucose chains from the shrinking solid substrate and thus expose fresh surfaces for hydrolytic attack.

6.4 ALKALI

Alkali treatment of cellulosics is the best known method of enhancement of biodigestion of cellulose. The application of alkali such as sodium

TABLE 6.1
A Quick Reference of Pretreatment Methods for Biological Methane Generation from Cellulosics

	Expected effects on				
Pretreatment	*Lignin seal*	*Cellulose crystal*	*Cellulose Hydrolysis rate*	*Cellulose Hydrolysis field*	*Cost and other general comments*
1. Size reduction milling	Rupture and increase in exposed ends	No decrystallization in general; ball milling decrystallizes	Enhancement of enzymatic hydrolysis expected	Increase not expected	High energy cost; ball milling may enhance both rate and yield
2. Alkali	Dissolution	Swelling	Expected enhancement for both acidic and enzymatic hydrolyses	Increase expected	Prior removal of hemicellulose preferred; otherwise high alkali cost due to acetyl groups
3. Sulfur dioxide	Dissolution	Little effect	Expected enhancement for enzymatic hydrolysis	Increase not expected	
4. Delignification by solvents	Dissolution	No effect	Expected enhancement for enzymatic hydrolysis	Increase not expected	Potential lignin byproducts
5. Biological delignification	Biodegradation	No effect	Expected enhancement for enzymatic hydrolysis	Decrease expected due to consumption of cellulose by the biological system	Too slow
6. Steam explosion	Rupture	Little effect	Expected enhancement for enzymatic hydrolysis	Increase documented for enzymatic hydrolysis but not for acid hydrolysis	Steam cost; loss of hemicellulose; potential lignin byproducts

7. Freeze explosion—liquid NH_3	Rupture	Decrystallization possible	Enhancement expected	Increase expected	Cost of NH_3
8. Mild acid hydrolysis	No effect	No effect	Expected enhancement in enzyme accessibility	Increase not expected	Low cost; recovery of hemicellulose sugars; residue easy to mill
9. Cellulose solvent	Separation, rupture	Decrystallization	Enhancement expected for both acidic and enzymatic hydrolyses	Complete conversion expected	High cost of solvent; recovery of solvent necessary, but difficult
10. Irradiation—gamma ray	Degradation due to free radicals	Not expected	Enhancement expected for enzymatic hydrolysis	Increase not expected	Radiation cost
11. Future methods—organic chemical treatments	Separation possible	Complete decrystallization possible	Enhance depolymerization rate	Increase in yield expected	May not yield glucose; many byproducts possible; may be uniquely suited for gaseous hydrocarbon generation but not for ethanol fermentation
12. Autohydrolysis	No effect	Decrystallization	Enhancement expected	Increase expected	
13. Enzymatic	No effect	Decrystallization	Enhanced rate	Increase in yield expected	High cost

hydroxide results in removal or rupture of lignin, an increase in surface area due to cellulose swelling, and also some degree of decrystallization of cellulose. The effectiveness of this pretreatment was demonstrated with an increased enzymatic digestibility by Guggolz *et al.*;[9] Han and Callihan;[10] Dunlop *et al.*;[2] and Destroy *et al.*[11] Jerger *et al.*[4] reported an enhanced rate of methane generation from three woody feeds: hybrid poplar, cottonwood, and sycamore. An increased methane yield was achieved from cottonwood, but not the other species. A concentration of 17 wt% NaOH was determined to be most effective toward increasing biodegradability. Chynoweth *et al.*[12] and Ghosh *et al.*[13] showed a 10–15% increase in methane yield from water hyacinth in both semicontinuous and batch-fed digester studies following alkaline treatment. Colleran *et al.*[14] examined a variety of pretreatment techniques and reported alkaline treatment to be the most effective for optimizing digester performance. A 50% increase in methane yield was achieved during digestion of wheat straw in a two-stage reactor system.

Alkali removes lignin and also 'swells' cellulose. What exactly takes place at a molecular level in swelling is still not clear. Under certain conditions, cellulose can be dissolved into a liquid form by alkaline solutions. The relative degree of swelling and dissolution is related to the extent of the treatment. At a molecular level, it would seem to be difficult to draw a demarcation where swelling ends and dissolution begins. Furthermore, one should also realize that recrystallization of cellulose after swelling and/or dissolution is a spontaneous process according to thermodynamic considerations. Therefore, from the viewpoint of process design, the time span between caustic swelling and the subsequent neutralization by acid followed by biological digestion would be of importance in deciding the precise enhancement in both the rate and the extent of methane generation from the pretreated substrate.

6.5 MILD ACID HYDROLYSIS AS A PRETREATMENT

Hemicellulose can be readily removed by dilute acid hydrolysis under gentle conditions to yield a hydrolysate containing pentoses, hexoses, and uronic acids. The hydrolysate should be a good liquid substrate for methane generation. A process involving the use of a minimum amount of dilute acid to achieve a high hydrolysate concentration of some 200 g of dissolved solids per liter has been developed by Tsao and Chiang.[3] The solid residue after removal of the solubilized hemicellulose contains mostly cellulose and lignin and is highly porous and weak in physical strength. The increased pore size, due to the removal of hemicellulose, can presumably increase the accessibility of cellulose to hydrolytic enzymes. The mild acid hydrolysis has therefore been looked upon as a pretreatment to enhance subsequent cellulose conversion.

While the Purdue process[3] involves using a mild acid as a pretreatment at a moderate temperature of 80–100 °C, a high-temperature dilute acid pretreatment is also reported by Knappert *et al.*[15] to remove hemicellulose. With a 0·2% sulfuric acid solution and a reaction time of 30 seconds at 180 °C, cellulosic material yields little glucose, indicating the limited degree of cellulose hydrolysis. This pretreatment procedure does, however, greatly reduce the polymerization of the cellulose, creating a residue that can then be hydrolyzed by enzymes. Residues from sulfuric acid treatment, which contain sulfates, can inhibit methane production by providing an alternative electron acceptor for the fermentation process.

Colberg *et al.*[18] reported that the heat treatment of biomass at temperatures of 175–225 °C for 1–2 hours caused the production of organic acids. These acids lowered the pH and resulted in increased rates of cellulose and hemicellulose hydrolysis to simple sugars and their dehydration products. Staged autohydrolysis solubilized over 40% of white fir organic matter and increased the biodegradability of this material to methane by almost 30%. Authohydrolysis at 150 °C improved the biodegradability of wheat straw and corn stover, but had an adverse effect at higher temperatures by converting biodegradable materials into refractory materials. Autohydrolysis liquors were fermented to methane in three different reactors, and a rotating biological disc and anaerobic filter operated reliably at a 1-day retention time.

6.6 SULFUR DIOXIDE

Data reported by Moore *et al.*[19] indicate that the lignin content of various woody materials decreased after sulfur dioxide treatment. They suggest that lignin depolymerization occurred first, followed by the treatment and removal of soluble products. The use of sulfur dioxide has advantages over alkali in the relative ease of its recovery and its low liquid requirement. Sulfur dioxide is thought to be effective mainly in lignin rupture and removal, even though a recent report by Connor[20] indicated that drastic reduction in cellulose polymerization occurs without changes in cellulose crystallinity or enzymatic digestibility.

Sulfur dioxide treatment can be considered a special form of pulping, which has the main objective of solubilizing and thus removing lignin, as well as some portions of hemicellulose, with a minimum alteration in cellulose. Such improved pulping methods as the sulfite process, involving a solution of calcium bisulfite containing excess sulfur dioxide, enable the production of pure cellulose with nearly complete lignin removal. However, conventional pulping processes are generally designed to produce a clean pulp product valued at a much higher price than that of methane. Consequently, economic considerations will rule out the possibility of applying conventional pulping methods as pretreatments for biological methane generation. On the other hand, it could be rewarding to reexamine the records of the not-so-successful developments of other pulping processes that could be very effective in lignin removal but are not acceptable for paper-making for other reasons.

6.7 LIGNIN REMOVAL BY SOLVENT EXTRACTION

Lignin is soluble in alcoholic solutions. Butanol and ethanol are the common alcohols useful for lignin removal. Bungay[21] outlined a process involving aqueous butanol to remove lignin with the residual cellulose reported to be 80–90% readily hydrolyzable with enzymes. Katzen *et al.*[22] have also described a pulping process using ethanol. Solvent extraction as a pretreatment does have the added potential benefit of obtaining a lignin coproduct that can presumably have a considerable market value.

Many other solvents have also been used for lignin removal and thus enhancement of cellulose digestion. Ethylene glycol was used by Pannir Selvam and Ghose[23] to improve hydrolysis of rice straw and sugar cane bagasse. Lignin can dissolve readily in ethylene glycol. Peracetic acid is another solvent that was used by Toyama[24] for the purpose of lignin removal. Sodium hypochlorite, being a strong oxidizing agent, can fragment lignin into soluble products and remove lignin by oxidative extraction.[25]

6.8 BIOLOGICAL PRETREATMENT

Suggestions have been made to use biological delignification as a pulping method. Such procedures could also be applied as a pretreatment for anaerobic methane generation. White-rot fungus, *Lentinus edodes*, for example, can reduce the lignin content of wood without rapid depolymerization of cellulose. Other reports on biological methods include those by

Brooks *et al.*[26] using a mold, *Phanerochaete chrysosporium*, for aerobic digestion of lignin, and by Ander and Eriksson[27] using a 'cellulase-less' mutant of *Polyporus adustus* and other white-rot fungi to remove lignin and hemicellulose.

Research on enzymatic pretreatment has been focused primarily on cellulose, and related to process development for enzymatic saccharification for biomass for production of liquid fuels. The principles, however, would be applicable to improvement of biodegradability for conversion to methane. The current status of this field was recently discussed in several reviews.[28–30] Several economic analyses have concluded that enzymatic processes are not yet economic without very cheap feedstocks or product subsidies. Research, however, should be continued in view of projected future scarcity of gas and other fossil fuels. Research is needed on the types of enzymes involved in separation and hydrolysis of biological polymers from the complex matrixes characteristic of biomass. The organisms and enzymes involved as well as factors limiting their activities should be investigated. Studies should include a search for new strains, improvement of existing strains, greater understanding of controls over enzyme production and secretion, and factors affecting production, secretion, stability, and activity. Finally, the further development of processes combining enzymatic with physical and chemical treatment is a frontier for research.

6.9 STEAM EXPLOSION

Pressuring wood with steam and then suddenly releasing it to atmospheric pressure can shatter the structure.[31] The shattered cellulosic biomass reportedly then becomes susceptible to hydrolysis. Steam explosion separates cellulose fibers in a popcorn-like manner due to fast evaporation of originally pressurized water in the biomass. The 205–260 °C temperature of the high-pressure steam degrades hemicellulose and causes liberation of organic acids from the plant material. These acids at the high temperature promote further degradation of the carbohydrates. Iotech Company in Canada has developed a process that involves a steam pressurization at 45–52 atmospheres followed by extrusion of the pressured biomass through heads to disintegrate the cellulosic material.[21] While 80% of the resulting cellulose is enzymatically digestible, the high steam temperature reduces the degradability of the hemicellulose fraction. The effect of steam explosion on acid hydrolysis has not been documented. If the explosion shatters the particle and the lignin seal without altering the cellulose crystallinity, enhanced acid hydrolysis of the pretreated cellulose may not be so obvious because the main resistance of cellulose to acid hydrolysis is due to the crystal structure.

6.10 FREEZE EXPLOSION

Liquid anhydrous ammonia has been shown to be a swelling agent for cellulose. Dale and Moreira[32] reported the development of a freeze-explosion technique for pretreating cellulose with liquid ammonia. Rapid depressurization causes fast evaporation of ammonia and thus shatters the solid substrate. The freezing temperature of the liquid ammonia eliminates the problem of heat and acid degradation of carbohydrates. The pretreated alfalfa and rice straw have produced greater than 90% of the theoretical maximum yield of glucose by enzymatic hydrolysis. The residual ammonia in the pretreated biomass can presumably provide an additional nitrogen source supporting microbial growth.

6.11 DISSOLUTION OF CELLULOSE BY SOLVENTS

Pretreatment by cellulose solvents can disintegrate the biomass structure by dissolving cellulose while rupturing the surrounding lignin barriers, thus removing the two major hindrances of cellulose hydrolysis in one operation.[33] Ladisch *et al.*[34] have reported quantitative yields of glucose from enzymatic hydrolysis of cellulose following pretreatment with cellulose solvents. The key finding was that cellulose need not be dissolved but merely soaked or permeated with solvent to disintegrate the crystalline structure. This could prove important in process development, since much less solvent is needed to soak cellulose than would be required to dissolve it to form a freely flowing liquid. Many cellulose solvents have been reported in the literature. Concentrated phosphoric acid, for example, was found by Walseth in 1952[35] to render cellulose highly reactive. Recently, solvents including Cadoxen, caustic ferric tartrate, concentrated sulfuric acid, and concentrated aqueous zinc chloride have been applied in pretreatment to achieve quantitative conversion of cellulose into glucose.[33,36–38]

Gaseous hydrogen chloride was also used in pretreatment of cellulosics. HCl, upon absorption by moisture normally existing in cellulosic materials, will form concentrated hydrochloric acid *in situ* that, at a 40% plus concentration, can also swell and dissolve cellulose. Although the exact function of the HCl in the so-called Japanese Chisso process of cellulose hydrolysis is not fully understood, it is likely that the cellulose-dissolving ability of this concentrated hydrochloric acid has a major role in achieving the greater than 95% cellulose conversion into glucose. The well-known Bergins-Rheinau process uses 40% liquid hydrochloric acid to first disrupt the cellulose structure to achieve high glucose yield. The commercial success of the HCl processes will require a nearly complete recovery and recycle of the fairly expensive hydrogen chloride. This, so far, appears to be difficult to achieve. The corrosive nature of the hydrochloric acid poses another technical problem.

6.12 IRRADIATION

Gamma radiation causes, in cellulosics: depolymerization of biopolymers, decomposition of carbohydrates, reduction in cellulose crystallinity, and increase in digestibility in subsequent hydrolysis by acid and enzymes.[39,40]

The effect of gamma radiation on the biodegradability of biomass and wastes is, however, affected by factors such as the availability of oxygen and feedstock composition. Gottschall *et al.*,[41] for example, digested sludge with a ^{60}Co source for 24 hours at a dose of approximately 0·4 mrad/h. They found that the treated sludge produced very little methane from the direct scission of organic material. Treatment of aqueous sludge suspensions with the same radiation dose in the presence of excess oxygen, however, gave consistent increases in gas yields of up to 50% during anaerobic digestion. Treatment in an air atmosphere showed a 16% increase in gas production, while treatment of sludge suspensions in a microaerophilic atmosphere decreased gas production by up to 26%. Other studies using aspen and spruce as feedstock demonstrated strong species specificity for the pretreatment of wood by irradiation. The *in vitro* digestibility of aspen increased to 78% while the increase in the digestibility of spruce was only 14% following an electron dosage of 108 rads. The digestibility of straw also improved.[42] This radiation offers a potentially useful pretreatment method. Large quantities of gamma-ray-emitting ^{137}Cs exist in fission wastes since the initiation of ^{239}Pu production. Pretreatment by irradiation could make use of the wastes and help reduce some of the disposal problems of radioactive wastes.

6.13 AUTOHYDROLYSIS

Autohydrolysis of biomass to increase its digestibility has been studied by McCarty *et al.*[43–45] This pretreatment technique involves thermal hydrolysis at 175–225 °C for the production of organic acids. The low pH then promotes the hydrolysis of polysaccharides to soluble products such as sugars, furans, and acetic acid. The first stage of autohydrolysis solubilizes the hemicellulose, while subsequent stages under harsher conditions solubilize cellulose, but also result in some sugar decomposition. This technique shows favorable results for increasing the biodegradability of white fir and wheat straw. However, at temperatures greater than 175 °C, the biodegradability of wheat straw decreases because soluble products formed during pretreatment are converted to nonbiodegradable residues.

Organic acids, mostly acetic acid, are released from plant materials when they are heated in the presence of water. These released acids can then promote additional hydrolysis. The mild acid hydrolysis, which can occur without the external addition of acid, is called autohydrolysis. Lora and Wayman[16] employed autohydrolysis to solubilize hemicellulose and to depolymerize lignin. In a more recent report by Linden *et al.*,[17] autohydrolysis of wheat straw at 180 °C for 45 minutes was found to enhance subsequent cellulose hydrolysis.

6.14 SUMMARY

Many of the current pretreatment methods are aimed at improved cellulose hydrolysis to produce fermentable sugars with increased yield and/or enhanced rate. Fermentable sugars can be fed to methane generators, and they can also be used as the primary carbon and energy sources for a host of fermentation processes to produce ethanol, organic acids, and other products. Cellulose hydrolysis to yield glucose is obviously not an easy task. With all the research and pretreatment methods developed in the past and in recent years, an unquestionably superior method of glucose production from cellulose has yet to emerge. Methane generation is carried out by microbial flora of mixed cultures of highly versatile biological capabilities; pretreatment to fermentable sugars may not be required to achieve good methane production. We thus suggest a broadened view of the processing of cellulosics specifically to optimize methane generation. We suggest a search for other potential pretreatment methods that may not necessarily enhance glucose production but some other intermediates, as long as these intermediates can be made efficiently from cellulosics and can be converted efficiently into methane.

Pretreatments often achieve cellulose decrystallization, rupture of lignin barriers, and/or fractionation of cellulose, hemicellulose and lignin into separate streams. A variety of process designs for biological methane generation are thus possible. A full-scale plant for methane generation from cellulosics may justify complicated process schemes that can include multiple processing steps and a number of co- and byproducts. A simple process design may involve no more than pretreatment of the biomass followed by feeding the whole pretreated biomass to an anaerobic digester. A somewhat more complicated version could include separation of lignin, feeding the residual cellulose and hemicellulose into anaerobic digestion, and processing the separated lignin. As cellulosic biomass becomes increasingly accepted in the process industries as a feasible source of raw materials, we must examine the potential for more sophisticated processing schemes with integrated production of a multitude of fuels and chemicals.

Cellulosics are generally resistant to hydrolysis by enzymes because of two major hindrances represented by the cellulose structure and the lignin barrier. Alkali treatment of cellulosics is the best known method of enhancement of biodigestion of cellulose. The application of alkali such as sodium hydroxide results in removal or rupture of lignin, an increase in surface area due to cellulose swelling, and also some degree of decrystallization of cellulose. This technique should be further examined on different biomass feeds to evaluate its effectiveness toward increasing methane production rates and methane yields.

The enhancement in biodigestibility of cellulosics by simple pretreatment of size reduction might be explained in terms of increased accessibility to ends of the long cellulose microfibrils wrapped inside lignin barriers, as well as the rupture of the barrier by the cutting action. Particle size reduction should be further investigated as a means of increasing methane yields and methane production rates from biomass feedstocks.

Lignin removal from native cellulosics has also been studied as a potential pretreatment technique for cellulose utilization. If, indeed, lignin contributes little as a carbon and hydrogen source for methane generation, and yet efficient methods are available for its separation from cellulose and hemicellulose, the need for an integrated scheme including utilization of lignin coproducts will arise for reasons of economics and process efficiency, as well as the alleviation of a waste disposal requirement that would otherwise exist and should be investigated.

New pretreatment methods for selective catalytic oxidation and chemical modification quantitatively convert cellulose into soluble, noncrystalline polymers. These polymers, being polyols, are biodigestible and can be fed into methane digesters. They can also be easily hydrolyzed into C_2 and C_4 fragments with or without simultaneous hydrogenation. The C_2 and C_4 fragments are again readily biodigestible. Pretreatment methods which should be further investigated include alkaline hydrogen peroxide treatment, hydrogen fluoride hydrolysis, wet oxidation, and ozonation.

The heating value of methane produced by anaerobic digestion is diluted by the coproduct carbon dioxide. Pretreatment by heating results in loss of oxygen in the form of water, production of less carbon dioxide, and thus enrichment of the methane content. Other methods of feed oxygen removal such as decarboxylation by wet carbonization should result in an enriched methane product gas.

REFERENCES

1. Millet, M. A., Baker, A. J. and Satter, L. D., 'Physical and Chemical Pretreatments for Enhancing Cellulose Saccharification,' *Biotech. Bioeng. Symp.* **6**:125–153, 1976.
2. Dunlop, C. E., Thompson, J. and Chiang, L. C., 'Treatment Processes to Increase Cellulose Microbial Digestibility,' *Am. Inst. Chem. Eng. Symp. Ser.* **72**:58–63, 1976.
3. Tsao, G. T. and Chiang, L. C., 'Cellulose and Hemicellulose Technology,' *The Filamentous Fungi* **4**:296–326, 1983.
4. Jerger, D. E., Dolenc, D. A. and Chynoweth, D. P., 'Biogasification of Woody Biomass Following Physical and Chemical Pretreatment,' Paper presented at *7th Symposium on Energy From Biomass and Wastes*, Orlando, Fla., 1983.
5. Pew, J. C. and Weyna, P., 'Fine Grinding, Enzyme Digestion, and the Lignin–Cellulose Bond in Wood,' *TAPPI* **45**:247, 1962.
6. Tassinari, T. and Macy, C., 'Differential Speed Two Roll Mill Treatment of Cellulosic Materials for Enzymatic Hydrolysis,' *Biotech. Bioeng.* **19**:1321–1330, 1977.
7. Nystrom, J., 'Discussion of Pretreatments to Enhance Enzymatic and Microbiological Attack of Cellulosic Materials,' *Biotech. Bioeng. Symp.* **5**:221–224, 1975.
8. Kelsey, R. G. and Shafizahed, F., 'Enhancement of Cellulose Accessibility

and Enzyme Hydrolysis by Simultaneous Wet Milling,' *Biotech. Bioeng.* **22**:1025–1036, 1980.

9. Guggolz, J., Kohler, G. O. and Klopfenstein, J. J., 'Composition and Improvement of Grass Straw for Ruminant Nutrition,' *J. of Animal Sci.* **33**:151–156, 1971.
10. Han, Y. W. and Callihan, C. D., 'Cellulose Fermentation: Effect of Substrate Pretreatment on Microbial Growth,' *Appl. Microbiol.* **27**:159–165, 1974.
11. Destroy, R. W., Lindenfelser, L. A., Sommer, S. and Orton, W. L., 'Bioconversion of Wheat Straw to Ethanol: Chemical Modification, Enzymatic Hydrolysis and Fermentation,' *Biotech. Bioeng.* **25**:1527–1535, 1981.
12. Chynoweth, D. P., Dolenc, D. A., Schwegler, B. and Reddy, K. R., 'Wastewater Reclamation and Methane Production Using Water Hyacinth and Anaerobic Digestion.' Paper presented at the 10th Energy Technology Conference, Washington, DC, 1983.
13. Ghosh, S., Henry, M. P. and Klass, D. L., 'Bioconversion of Water Hyacinth-Coastal Bermuda Grass-MSW-Sludge Blends to Methane,' *Biotechnol. Bioeng. Symp.* **10**:163–87, 1980.
14. Colleran, E., Barry, M. and Wilkie, A., 'The Application of the Anaerobic Filter Design to Biogas Production From Solid and Liquid Agricultural Wastes,' pp. 443–482 in Symposium Papers, *Energy From Biomass and Wastes VI*. Chicago: Institute of Gas Technology, 1982.
15. Knappert, D., Grethlein, H. and Converse, A., 'Partial Hydrolysis of Cellulosic Materials as a Pretreatment for Enzymatic Hydrolysis.' Paper presented at the *American Institute of Chemical Engineers National Meeting*, San Francisco, 1979.
16. Lora, J. H. and Wayman, M., 'Delignification of Hardwood by Autohydrolysis and Extraction,' *TAPPI* **61**:47–50, 1978.
17. Linden, J. C., Murphy, V. G. and Moreira, A. R., 'Wheat Straw Autohydrolysis,' *Adv. in Biotech.* **2**:41–46, 1980.
18. Colberg, P. S. *et al.* 'Heat Treatment of Organics for Increasing Anaerobic Biodegradability,' *Annual Progress Report*, SERI/TR-98174-1, March 1981.
19. Moore, W. E., Effland, M. J. and Millet, M. A., 'Hydrolysis of Wood and Cellulose With Cellulolytic Enzymes,' *J. of Agricultural and Food Chemistry* **20**:1173–1175, 1972.
20. Connor, A. H., 'Effect of Aqueous Sulfur Dioxide on Cellulose,' *Biotechnology Letters* **2**:439–444, 1980.
21. Bungay, H. R., *Energy, The Biomass Option*. New York: Wiley, 1981.
22. Katzen, R., Frederickson, R. and Bush, B., 'The Alcohol Pulping and Recovery Process,' *Chem. Eng. Progr.* **76**:62–67, 1980.
23. Pannir Selvam, P. V. and Ghose, T. K., 'Pretreatment of Agricultural Residues for Enzymatic Saccharification by the Culture Filtrate of *T. reesei* QM914.' Paper presented at the 2nd International Conference on Bioconversion and Biochemistry, Eng., IIT, Delhi, 1980.
24. Toyama, N., 'Feasibility of Sugar Production From Agricultural and Urban Cellulosic Wastes With *Trichoderma viride* Cellulase,' *Biotech. Bioeng. Symp.* **6**:207–219, 1976.
25. Fan, L. T., Gharpuray, M. M. and Lee, Y. H., 'Evaluation of Pretreatments

for Enzymatic Conversion of Agricultural Residues,' *Biotech. Bioeng. Symp.* **11**:29–45, 1981.
26. Brooks, R. E., Bellamy, W. and Su, T., 'Bioconversion of Plant Biomass to Ethanol,' NTIS, COO/4147/4, 1978.
27. Ander, P. and Eriksson, K. E., 'Lignin Degradation and Utilization by Microorganisms,' *Progr. Indust. Microbiol.* **14**:1–58, 1978.
28. Mandels, M., 'Cellulases', in G. T. Tsao, Ed., *Annual Reports on Fermentation Processes*, Vol. 5. New York: Academic Press, 1982.
29. Millet, M. A., Baker, A. J. and Satter, L. D., 'Pretreatment to Enhance Chemical Enzymatic and Microbiological Attack of Cellulosic Materials,' *Biotechnol. Bioeng. Symp.* **5**:193–219, 1975.
30. Spano, L., Tassinari, T., Ryn, D., Allen, A. and Mandels, M., 'Producing Ethanol From Cellulosic Biomass,' in *Proc. of Symp. on Biogas and Alcohol Fuels Production*, Emmaus, Pa.: J. G. Press, 1980.
31. Cahal, D. S., McGuire, S., Pikor, H. and Noble, G., 'Bioconversion of Hemicellulose and Cellulose of Wood into Fungal Food and Feed With *Chaetomium cellulolyticum.*' Paper presented at the 2nd World Congress of Chemical Energy, Montreal, 1981.
32. Dale, B. E. and Moreira, M. J., 'A Freeze-Explosion Technique for Increasing Cellulose Hydrolysis,' *Biotech. Bioeng. Symp.* **12**:31–43, 1982.
33. Tsao, G. T., Ladisch, M., Ladisch, C., Hsu, T. A., Dale, B. and Chou, T., 'Fermentation Substrates From Cellulosic Materials: Production of Fermentable Sugars From Cellulosic Materials,' *Annual Reports Ferment. Processes* **2**:1–22, 1978.
34. Ladisch, M. R., Ladisch, C. and Tsao, G. T., 'Cellulose to Sugars: New Path Gives Quantitative Yield,' *Science* **201**:743–745, 1978.
35. Walseth, C. C., 'The Influence of the Fine Structure of Cellulose on the Action of Cellulases,' *TAPPI* **35**:233–238, 1952.
36. Tanaka, M., Taniguchi, M., Morita, T., Motsuno, R. and Kamikubo, T., 'Effect of Chemical Treatment on Solubilization of Crystalline Cellulose and Cellulosic Wastes With *Pellicularia filamentosa* Cellulase,' *J. of Fermentation Technol.* **53**:186–190, 1979.
37. Sasaki, T., Tanaka, T., Nanbn, N., Sato, U. and Kainuma, K., 'Correlation Between X-Ray Diffraction Measurement of Cellulose Crystalline Structure and the Susceptibility to Microbial Cellulose,' *Biotech. Bioeng.* **21**:1031–1042, 1979.
38. Chen, L. F. and Gong, C. S., 'Enzymatic Hydrolysis of Cellulose Pretreated With Zinc Chloride and Hydrochloric Acid,' *Biotech. Bioeng. Symp.* **12**:57–65, 1982.
39. Han, Y. W. and Ciegler, A., 'Effect of Gamma-Ray Irradiation on Sugar Production From Plant Biomass,' *Biotech. Bioeng. Symp.* **12**:73–77, 1982.
40. Han, Y. W. and Ciegler, A., 'Effect of High Energy Radiation on Lignocellulose Conversion,' *Annual Reports on Fermentation Processes* **6**:299–322, 1983.
41. Gottschall, W. C., Morse, J. G., Updegraff, D. M. and Wang, A., 'The Effect of Gamma Radiation on Yields of Methane Gas From the Anaerobic Bacterial Digestion of Sewage Sludge,' pp. 193–214 in Symposium Papers, *Energy From Biomass and Wastes III*. Chicago: Institute of Gas Technology, 1978.

42. Millet, M. A., Baker, A. J. and Satter, L. D., 'Pretreatments to Enhance Chemical, Enzymatic and Microbiological Attack of Cellulosic Materials,' *Biotechnol. Bioeng. Symp.* **10**:93, 1980.
43. McCarty, P. L., Young, L. Y., Stuckey, D. C. and Healy, J. B., 'Heat Treatment for Increasing Methane Yields From Organic Materials,' pp. 179–99 in Schlegel, H. G. and Barnea, J., Eds., *Microbiology Energy Conversion*. New York and Oxford: Pergamon Press, 1977.
44. Colberg, P. J., Baugh, K., Everhart, T., Bachman, A., Harrison, D., Young, L. Y. and McCarty, P. L., 'Heat Treatment of Organics for Increasing Anaerobic Biodegradability.' Annual Progress Report for June 1979–August 1980 to Solar Energy Research Institute, SERI/TR-98174-1, March 1981.
45. McCarty, P. L., Baugh, K., Bachmann, A., Beard, V. and Levy, J., 'Autohydrolysis of Organic Residues to Increase Biodegradability to Methane,' in *Fuel Gas Production From Animal and Agricultural Residues and Biomass: Proceedings of the 1982 Anaerobic Digestion Program Review Meeting*, SERI/CP-231-1924, DE83011965, March 1983.

7
Nutrient Requirements

RICHARD E. SPEECE
Environmental Studies Institute, Drexel University, Philadelphia, Pennsylvania, USA

7.1 INTRODUCTION

Our lack of knowledge of the nutritional requirements of methanogens, in particular for nickel, selenite, and molybdate, has hindered development of the anaerobic digestion process for decades. Even though iron and cobalt requirements were reported on by Speece and McCarty,[1] there is evidence that these elements are still limiting for some substrates.[2,3] The first substrates widely used for the anaerobic digestion process were domestic wastewater sludges. Such sludges have generally been assumed to contain all of the required nutrients, in concentrations adequate to prevent nutrient limitations on the anaerobic digestion process. However, results reported by Speece[4] indicate that there is some preliminary evidence that iron availability may be limiting methanogenic activity even with domestic wastewater sludge as a feedstock.

As the anaerobic digestion process is applied to other feedstocks, the nutritional requirements will demand increasing attention. Some candidate biomass feedstocks under consideration for conversion to methane such as corn, sugarcane, sorghum, and wood appear to be nutrient-limited based on elemental carbon, nitrogen, and phosphorus data and laboratory digestion data, indicating that nutrient supplementation may be necessary for biogasification.[5,6] Other feeds such as kelp and water hyacinth vary from nutrient-rich to nutrient-limited depending upon growth conditions.[6,7] Furthermore, requirements for micronutrients for anaerobic digestion of biomass are unknown. The trend seems apparent; the diversity of components of these feedstocks is more limited than in domestic wastewater sludges. It is logical to extrapolate that the required nutrients would also be more limited than in domestic wastewater sludges. This progressive limitation of inherent nutrients is further manifested with many industrial wastewaters that are candidate feedstocks for anaerobic digestion.

7.2 NUTRITION AND GROWTH

Microbial regeneration time is a function of the concentration of nutrients present. Ideal nutrient concentrations are not essential, but process compensation must be made in either lower loading rates or lower resulting treatment efficiency as a trade-off for non-ideality of chemical nutrients present. Nutrient limitations at best prolong the minimum generation time of methanogens and at worst can cause generation eventually to cease altogether. The rates of substrate metabolism are also limited by nutrient limitations. Specific substrate utilization rates can be increased several-fold when all required nutrients are in excess. There is a possibility that some of the published kinetic coefficients may reflect nutrient-limiting conditions under which the data were taken — for example, nickel and iron deficiency. In addition, toxicity response can be caused or compounded by nutrient limitations. For example, nitrogen-deficient kelp resulted in digester imbalance presumably because of the selective inhibition of methanogenic bacteria.[5]

A nutrient limitation will result in decreased rate of growth. The search for possible stimulants is therefore very dependent upon the nutrient medium used and is further complicated by the ecological interactions that must be taken into account. The requirements for nutrient supplementation in nutrient-deficient feeds such as wood and municipal solid wastes (MSW) may be successfully met by blending of nutrient-rich feeds such as domestic sludge.[8]

The literature contains very little information on the nutrient requirements of the hydrolytic, acid fermenting, or acetogenic bacteria. Relatively more information is reported on the nutrient requirements of the methanogens. Nutritional deficiency in the acetogens, hydrogen-utilizing methanogens, or acetate-utilizing methanogens could contribute to elevated volatile acids concentrations.

The nutritional stimulation of methanogens has been reported by Speece.[4] The nutrients, in decreasing order of importance, are: nitrogen, sulfur, phosphorus, iron, cobalt, nickel, molybdenum, selenium, riboflavin and vitamin B_{12}. Methane fermentation will occur without all of the nutrients present but the rates will be reduced. All of the nutrients of greater importance must be present before a lesser required nutrient will demonstrate stimulation. Without adequate nitrogen, sulfur will not show stimulation and without adequate sulfur, iron will not show stimulation. Ammonium appears to serve as the nitrogen source for all methanogens. Most methanogens utilize sulfide as a sulfur source, but some can utilize cysteine.[9]

A compound stimulatory to a pure species of bacteria may have no effect on a mixed culture such as found in an anaerobic digester. Mah *et al.*[10] found the rate of methane production differed in pure culture versus an enriched culture and stated:

> The interactions demonstrated between methanogenic and non-methanogenic species show that methane production in the mineral acetate enrichment is not the function solely of the organism catalyzing the split of acetate to methane and carbon dioxide; it is a function of a community of organisms, each contributing nutrients to the common environment, and withdrawing others. . . . Decomposition of acetate to methane occurs far more rapidly in the community of mixed species than with the sarcina alone. This increase in rate through the interaction of many microbes is the most significant feature of these studies insofar as practical problems of methanogenesis are concerned. Complex organic growth factors are released into the environment common to all the cells, hastening their growth and the rate of substrate disappearance. If provision of increased complex nutrients is the explanation for the increased methanogenesis by the consortium, addition of the appropriate complex nutrient substrates to the enrichment ecosystem should increase the rate of methane formation. The greater rate of methanogenesis by the sarcina in the complex medium as compared to acetate alone is consistent with this view.

Nutritional requirements are often reported for batch growth systems. However, there are severe limitations with the extrapolation of batch growth nutritional requirements to continuous culture systems typical of prototype biomass digesters. The continuous culture systems are very dynamic in nature:

(1) Shifts in organism predominance have been observed to occur cyclically.
(2) The organisms cycle through different growth stages.

(3) The aquatic chemistry of the system continuously varies with unsteady-state levels of trace metals and sulfide.
(4) It is presumed that the trace organics also fluctuate.
(5) The activity of the sulfate reducers fluctuates. Therefore, since meaningful extrapolation of nutrient requirements derived from batch growth systems to continuous culture systems is questionable, only continuous culture systems should be assayed for nutrient requirements.

Process optimization seeks to increase the rate-limiting step. Nutrient limitation may restrict this rate-limiting step in two ways. It could curtail the rate of activity of the class of microorganisms responsible for the rate-limiting step in the overall conversion, or it could preclude the predominance of a more efficient, higher rate class of microorganisms responsible for this step.

The unit rate of activity of any class of microorganisms in the anaerobic digestion process appears to be a multiplicative function related to the concentration of all required nutrients. Absence of any one essential nutrient can stop the process. However, the literature does not report the half-saturation concentration coefficient (K_S) for any of the inorganic nutrients — nitrogen, phosphorus, sulfide, iron, cobalt, nickel, molybdenum, tungsten or selenium. Also, K_S is not reported for any trace organic nutrients either. This has a strong economic impact, because, with many candidate biomass feedstocks, nutrient supplementation is required, imposing an economic burden on the process. Excess supplementation of a nutrient is to be avoided, but the K_S curve must be defined to optimize the minimum concentration of any required nutrient. It would be helpful to develop a model based upon chemical costs of each nutrient for process optimization of the rate and amount of methane production. This would help to define the practical trade-off between sacrifice of process efficiency and cost of chemical supplementation that may possibly arise with some of the more expensive trace organic stimulants.

7.3 NITROGEN AND PHOSPHORUS

Generally, nitrogen is required as the major nutrient other than an energy source for microbial systems. Speece and McCarty[1] determined the nitrogen requirements for the anaerobic digestion of fatty acids, carbohydrates, and proteins. The following growth equations were derived:

Amino and fatty acids	$A = 0{\cdot}054 - 0{\cdot}03\, M/F$
Glucose and starch	$A = 0{\cdot}46 - 0{\cdot}088\, M/F$
Nutrient broth	$A = 0{\cdot}076 - 0{\cdot}014\, M/F$

where A = fraction of substrate synthesized into biomass; F = mass of substrate metabolized by biomass per day measured as COD; M = mass of biomass in system.

The empirical formula for the anaerobic microbial cells was found to be $C_5H_7O_3N$, which has an N/cell ratio of 11% (w/w). At a 5% net cell synthesis ratio, the nitrogen requirement is 6 kg N/1000 kg of COD or 1 kg N/60 m^3 of methane produced. The empirical formulation of aerobic microbial cells was reported to be $C_5H_8O_2N$ by Symons and McKinney[11] while Hoover and Porges[12] reported $C_5H_7O_2N$.

It is noted that microbial synthesis — and thus the nitrogen requirement of carbohydrate digestion — is about six times greater than for proteins and fatty acids, respectively. This is shown in Fig. 7.1. Therefore, the anaerobic digestion of high carbohydrate feedstocks to methane gas deserves special consideration because of the relatively high microbial-synthesis-to-substrate ratio.

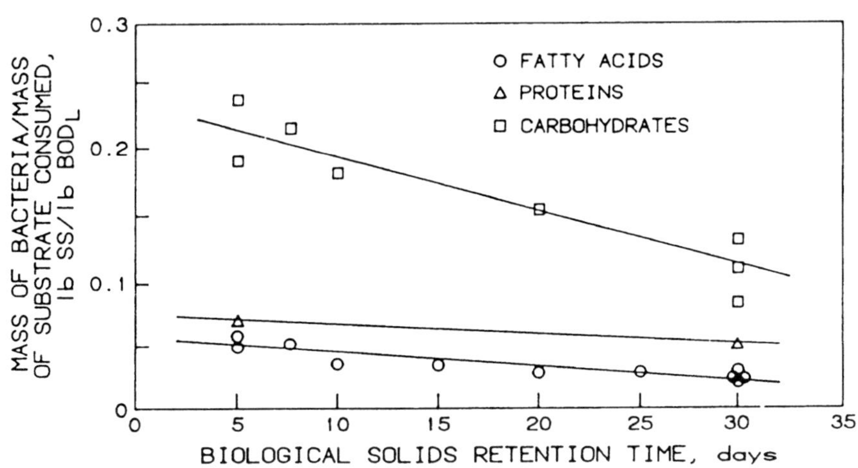

FIG. 7.1. Biological solids production resulting from methane fermentation.[4]

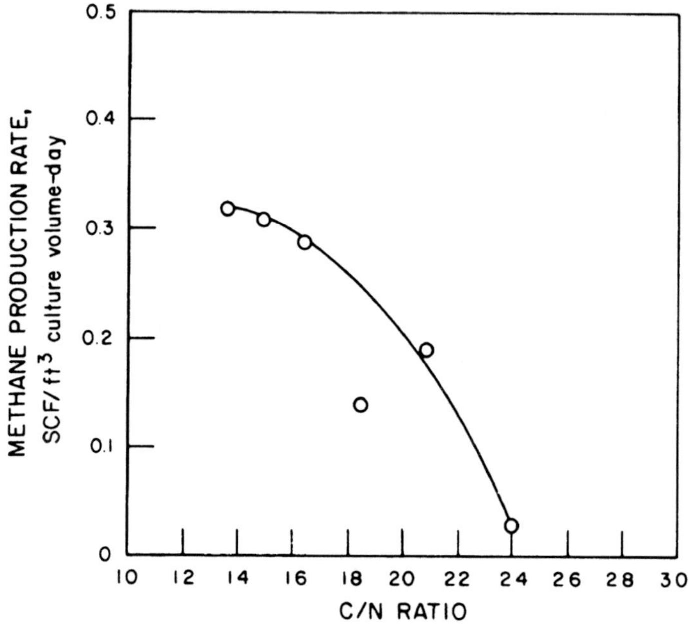

FIG. 7.2. The effect of nitrogen supplementation (NH_4Cl) in anaerobic digestion of kelp[5] (STR, loading—0·1 lb VS/ft^3-day, retention time—15 days, 35 °C).

A carbon-to-nitrogen ratio of 15:1 and a carbon-to-phosphorus ratio of 75:1 was recommended for the non-nutrient-limited digestion of kelp in stirred tank reactors operated at a 15–18-day SRT.[5] Other investigators have indicated that a carbon-to-nitrogen ratio of 23:1 and a carbon-to-phosphorus ratio of about 113:1 are needed in fixed-film reactors.[13]

The relationship between C/N and methane production from kelp is illustrated in Fig. 7.2.[5] It should be emphasized that the biodegradable values of N and P actually determine limiting conditions. However, total values for these elements such as those shown in Table 7.1 for various biomass feeds provide a rough estimate of potential nutrient limitations.[5] These data suggest that many types of biomass feeds may require supplementary nutrients for bioconversion. The nutrient content of biomass is variable within a given species depending upon growth conditions. The actual requirements, however, are based on the biologically available nutrients.

Reactors that promote longer SRTs have lower cell synthesis requirements and, consequently, have lower energy and nutrient requirements.

TABLE 7.1
Nutrient Ratios of Biomass and Waste Feedstocks*

Feedstock	*Carbon/nitrogen*	*Carbon/phosphorus*
Aquatic		
Kelp	15	84
Water hyacinth	10	94
Herbaceous		
Bermuda grass	40	194
Napier grass	41	527
Cattail	41	278
Woody		
Eucalyptus	490	446
Sycamore	178	2 480
Loblolly pine	432	2 600
Wastes		
Primary sludge	8	50
Municipal solid waste	76	204

*Data from Ref. 5.

The ammonia nitrogen concentration in the effluent from kelp-fed nonmixed vertical flow reactor (NMVFR) and continuously fed stirred tank reactor (CSTR) digesters operated at 35 °C and at SRTs ranging from 7 to 200 days and at loading rates ranging from 0·4 to 9·6 kg/m^3-day is illustrated in Fig. 7.3. The ammonia nitrogen concentration in the effluent of a stirred tank reactor operated at an SRT of about 7 days was about 110–220 mg/liter. In a NMVFR operated at an SRT of about 66 days, however, the ammonia nitrogen concentration was about 750–770 mg/liter. And the digester operating at an SRT of approximately 200 days had an ammonia nitrogen concentration of about 790 to 810 mg/liter in the effluent.[7]

These data, from several digesters operated at different SRTs, support the hypothesis that the nitrogen requirements for anaerobic digestion can be reduced by increasing the SRT. They further suggest that the limiting carbon-to-nitrogen ratio may be higher in reactors that promote longer SRTs and that the ammonia-enriched effluent from such reactors will have a higher value as a potential fertilizer.

The microbial uptake of phosphorus in anaerobic digestion has been reported to be approximately 14% of that for nitrogen.[1] This would yield

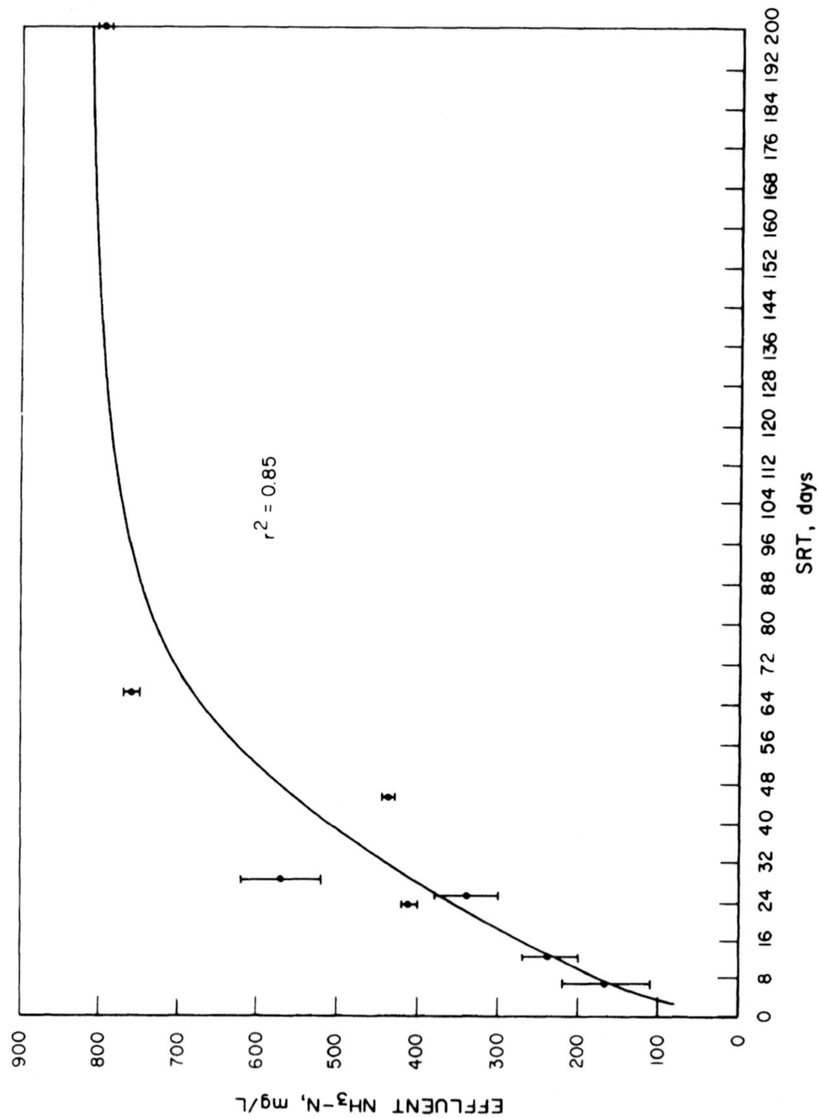

FIG. 7.3. Relationship between SRT and ammonia nitrogen in kelp digester effluent.[7]

an empirical microbial formulation of $C_5H_7O_2NP_{0.06}$. No data on the half-saturation constant for phosphorus is reported in the literature, but it is assumed to be on the order of 1 mg/liter as P or less.

7.4 SULFUR

The sulfur requirement for methanogens is not extensively documented. The limited references in the literature make reference to only two of the many methanogenic pure cultures and one mixed culture, which had cellulose as the substrate.[14–20] The sulfur requirement is part of a complex picture. Although the presence of sulfate may limit methanogenesis because the sulfate reducers have a lower half-saturation coefficient than the methanogens for substrates such as hydrogen and acetate, methanogens are dependent on the production of sulfide for growth. It is known that methanogenic bacteria require fully reduced sulfur as a sulfur source.[21–23]

Data from Ronnow and Gunnarsson[16] showed that when sulfide was added in increments of 20 mg/liter, the methane production rate increased until sulfide was completely depleted in the medium (as determined by sulfide analysis) and that the methane production rate then decreased. Growth continued to increase during sulfide starvation.

A sulfur source of about 0·85 mM was found to be essential for degradation of cellulose to methane.[19] At 9 mM, all inorganic sulfur compounds other than sulfate inhibited both cellulose degradation and methane formation. The inhibition increased in the following order: thiosulfate $<$ sulfite $<$ sulfide $<$ H_2S.

Sulfide was found to act as a sulfur source rather than as a reducing agent.[20] The optimum sulfide concentration was found to be 10^{-4}M which coincides with the sulfide concentration in the rumen. In the absence of sulfide, cysteine was the only compound that stimulated methanogenesis.

The sulfide requirements can be easily overlooked in anaerobic digestion because there are many sinks for sulfide:

(1) H_2S in the off-gas.
(2) Total sulfides in the effluent.
(3) Microbial synthesis of sulfide (150% of P synthesis).
(4) Precipitation of sulfides by heavy metals.

One could reasonably hypothesize from these references that about 12 mg/liter (corresponds to 0·5% H_2S in gas) of unionized H_2S are

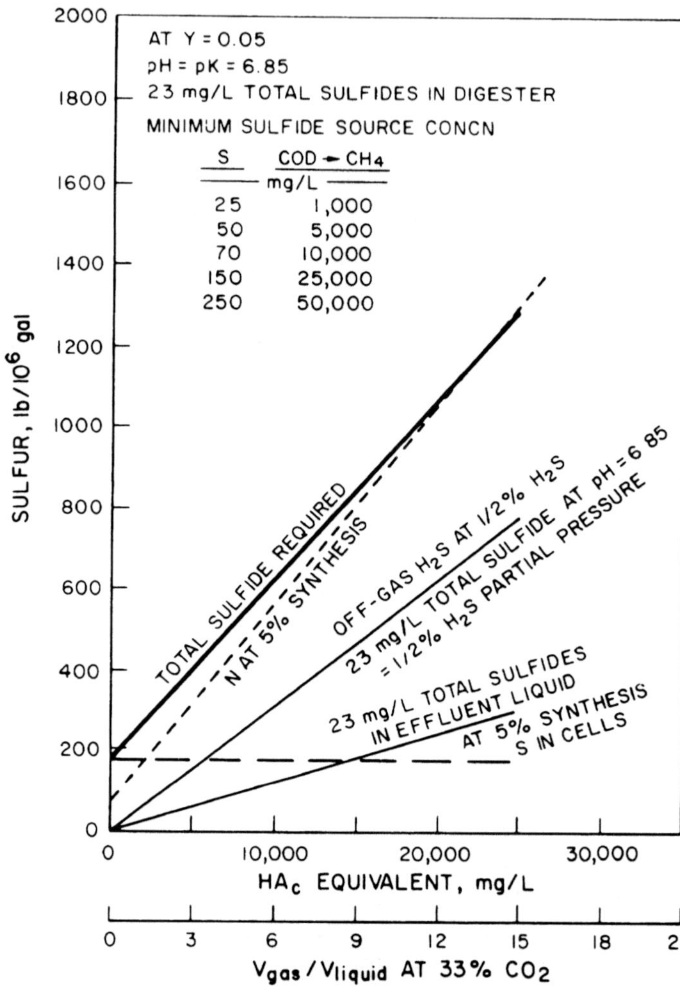

FIG. 7.4. Sulfur requirement for anaerobic digestion.

required for optimum growth of methanogens. If it were also assumed, for convenience, that the digester was operated at a pH of 6·85, equal to the pK of H_2S, then the total sulfide requirement for anaerobic digestion could be calculated as shown in Fig. 7.4.

To maintain 11·5 mg/liter of unionized H_2S in solution at a digester pH = pK of H_2S would result in a total dissolved sulfide concentration of 23 mg/liter, which would be lost in the effluent of a CSTR. A concentration

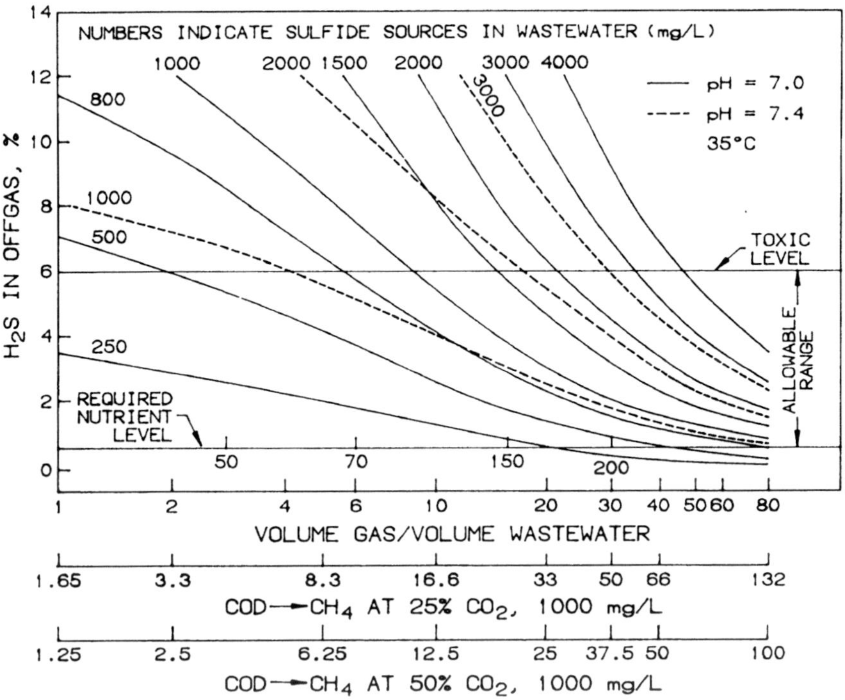

FIG. 7.5. H_2S in offgas versus V_g/V_w (or COD → CH_4).

of 11·5 mg/liter of unionized H_2S in solution would be in equilibrium with about 0·5% of H_2S in the gas phase. The mass of H_2S stripped into the gas phase is related to the volumes of digester gas produced per volume of liquid feed to the digester, and is shown in Fig. 7.5. The net result of this hypothesis is that the total sulfide requirement for optimum anaerobic digestion is essentially equivalent to the nitrogen requirement if a CSTR configuration of a digester is used. This is quite surprising since it is generally considered that the nitrogen requirement for anaerobic digestion is about an order of magnitude higher than any other nutrient other than the energy source. The second surprising observation is that the sulfur concentration of methanogenic cells is about 150% greater than that of phosphorus (phosphorus is normally assumed to be second only to nitrogen).

If it is assumed that both the required nutrient form and also the toxic form of sulfide is the unionized sulfide form, then Fig. 7.4 can be constructed and the ionized sulfide concentrations can be ignored in our range of interest. The toxic level of total dissolved sulfide in anaerobic digestion has been reported to be 200–300 mg/liter.[24] In such a case, if the digester pH is assumed to be 6·85 (the pK for H_2S) then the toxic concentration of unionized H_2S would be 100–150 mg/liter. Other investigators have reported that approximately 6% H_2S in the head gas is the maximum limit for methanogenesis.[25] This would correspond to 138 mg/liter of unionized H_2S in solution, which reasonably confirms the results of Lawrence and McCarty.[26]

Consequently, it could be concluded that approximately 0·5% H_2S is required in the head gas for optimum anaerobic digestion — 11·5 mg/liter unionized H_2S in solution (0·5% amounts to 6 times the H_2S in digester gas allowable by internal combustion engine warranties, which is 0·115 kg $H_2S/100\,m^3$ of digester gas). If 6% H_2S is the maximum toxic level for methanogenesis, then this illustrates the relatively tight ecological niche occupied by methanogens — at minimum desirable levels of approximately 0·5% and a maximum toxic level of 6% H_2S. Few inorganics go from ideal to toxic within such a small range as the 12-fold concentration hypothesized for sulfide.

7.5 TRACE METALS

The complex chemical environment in an anaerobic digester makes it difficult to predict the equilibrium concentration of heavy metals available in solution for microbial nutrition. Although investigators report on the concentration of minerals added to the nutrient media, the concentration of each metal in solution during digestion is rarely reported, even though this is more significant than the concentration added. The bicarbonate, phosphate, and sulfide ions affect precipitation and availability of nutrients.

Even measurement of trace metals in solution is not a true measure of their availability to the microorganism. The metals may be chelated so strongly that bacteria are unable to utilize them. Applying this principle, EDTA is used as a preservative because it ties up the metals so strongly that the bacteria are depleted of required metals for growth.

The chemical environment within an anaerobic digester apparently is mutually exclusive for the required trace metals and sulfide. According to known chemical laws, they cannot coexist in soluble forms in the required concentrations. However, the fact that both trace metals and sulfide do coexist in a digester in soluble form indicates that the chemistry must be complex and not understood.

Bacteria produce substances called siderophores and enterochelins that capture and transport trace metals into the cell. These substances may be responsible for the observed concentrations of one or more mg/liter of iron in solution (as measured by atomic absorption spectroscopy) simultaneously with measurable concentrations of hydrogen sulfide. The chemistry and chelation characteristics of anaerobic digestion environments deserve special attention because they bear so strongly on nutrient availability for microorganisms.

Nickel is generally not considered essential for growth of bacteria.[27] However, in the last few years it has become evident that the methanogenic bacteria are clearly a unique group in this respect. Nickel is an essential component of bacterial ureases and in bacterial enzymes which convert H_2–CO_2 to acetate.

In view of the apparently universal occurrence of F_{430} in methanogens and the inherent nickel requirement of F_{430}, the question arises as to the source of nickel, since it has not generally been included in defined media. The medium described by Balch *et al.*[28] for growth of most methanogenic bacteria contains 0·2% yeast extract and 0·2% trypticase, but it is not supplemented with nickel salts. Diekert *et al.*[29] found that the methanogenic bacteria tested grew readily on this medium. However, if the concentration of yeast extract was significantly lowered, good growth was only observed upon addition of nickel. They found that no other transition metal could substitute for nickel. They concluded that the high contamination level of nickel in the yeast extract medium is the reason why the nickel requirement for growth of methanogenic bacteria has long been overlooked. Mah *et al.*[30] showed that the ash of yeast extract was approximately as stimulatory as yeast extract itself, indicating that trace minerals, limiting in the defined medium, were supplied by yeast extract. They ruled out the possibility that the trace metal could be cobalt. In addition, the mineral salts and trace elements used to make the medium are contaminated with low amounts of nickel.[31] The nickel concentration in the medium prepared by the method of Balch *et al.*[28] was found to contain between 0·1 and 0·5 μM, even though nickel was not added. Diekert *et al.*[29] estimated the maximal amount of nickel required for growth of different methanogens to be 250–1100 μM of nickel per gram (dry weight) of cells.

Other trace metal requirements have been demonstrated with methanogenic bacteria. Iron, cobalt, molybdenum, selenium, and tungsten have been shown to be stimulatory to methanogens.[32] However, it appears that nickel is essential and that iron, cobalt or molybdenum cannot replace the nickel requirements for F_{430}.[33] Thus, there is clear documentation of the essential role of nickel as a nutrient requirement in anaerobic digestion.

The methanogens require seemingly contradictory growth conditions. For good growth, a sulfide concentration of 10–27 mg/liter is required.[14,15] But sulfide may remove metal ions essential for growth, such as nickel, iron, and cobalt.[34,35] Likewise high rates of gas production tend to strip essential sulfide from solution.[24] In one study the sulfur source was sulfate, which apparently was continuously reduced to sulfide. This provided a continual 'source' of sulfide to replenish that stripped by the high gas production rates. In addition the trace metals nickel, iron, and cobalt were fed in a 50% acetic acid feed with a low pH to keep the metals in solution. The pH controller automatically injected 1–2 milliliters of this acidified feed containing the soluble trace metals once per 15 or 30 minutes when the acetate-utilization rate was 25 g/liter-day.[36]

Speece and McCarty[1] indicated that even though sulfide would precipitate nickel, iron, and cobalt, thus leaving very low concentrations of the metals at equilibrium, the equilibrium was frequently disturbed by the acidified feed pulses making the trace metals momentarily available for uptake by the methanogens.[36] Thus nutrient limitation in the methanogens could be minimized, and the acetate energy source could be assayed as the limiting nutrient. It is not known if all nutrients were in fact unlimiting but exceptionally high acetate-utilization rates were observed, in the order of 15 times greater than that which is conventionally considered to be a high rate of digestion: that is, 3·3 g/liter-day. This is quite significant since enrichment cultures generally demonstrate lower acetate-utilization rates than normal sludge digesters.

It also appears that nickel, iron, and/or cobalt must be supplemented to achieve high microorganism (VSS) concentrations.[36] With yeast extract supplementation only, the highest volatile suspended solids concentration observed was 3000 mg/liter. When nickel and yeast extract were both supplemented, the highest VSS observed was 7000 mg/liter.

7.6 ECONOMICS

Economics will often dictate which nutrients can be justified for achievement of stimulated activity. In the conversion of acetate to methane, approximately 3–6 kg of nitrogen are required per 1000 kg of acetate utilized, depending on the SRT.[1] For a completely nitrogen-deficient fatty acid wastewater, if supplemental nitrogen costs \$0·40/kg (as N) and methane has a value of \$4·50/10^6 Btu, the nitrogen supplementation costs would be approximately 2% of the value of the methane produced. The phosphorus requirement is approximately 0·5 kg/1000 kg of acetate utilized.

Heavy metals are relatively inexpensive to supplement. Due to the presence (and requirement) of sulfide, most of the heavy metals are precipitated from solution and it is difficult to determine their actual requirement. Generally for methanogens, iron is added at approximately 10 mg/liter, cobalt at 5 mg/liter and nickel, molybdenum, and selenium at 0·1 mg/liter.

7.7 ORGANIC GROWTH FACTORS

Trace organic nutrient requirements for anaerobic digestion are not well defined. The success of 'reinoculation' from an active digester to stimulate a digester declining in activity implicates the requirements for trace organics.[4] Also, the more robust growth observed in mixed cultures versus pure cultures is another implication of trace organic stimulation. Various trace organics have been shown to stimulate methanogens. Among these are thiamine, vitamin B_{12}, riboflavin, hemoglobin, benzimidazole, proline, and the complex organic sources yeast extract, trypticase, Milorganite, etc. The relatively high cost of most of these trace organics would limit their practical supplementation to commercial digesters. However, the stimulation response curve should be defined for trace organics.

Organic growth factors which may be required for optimum activity of methanogens may include: coenzyme M, Factor 420, acetate, 2-methyl butyric acid, vitamins, N-acetyl glucosamine, riboflavin, B_{12}. Tanner[37] developed a defined medium for *Methanomicrobium mobile*. This organism was found to require acetate, isobutyrate, 2-methylbutyrate, isovalerate, tryptophan, para-aminobenzoic acid, pyridoxine, thiamine, biotin, and vitamin B_{12} for its growth. ATP is also required during the methane formation process.[38–40] Several of these compounds are only required by particular species. CoM and F_{420} have been discussed above. Trace amounts of nickel are stimulatory to an enriched culture of acetate-utilizing methanogens. Diekert *et al.*[27,33,41] found nickel to be a component of F_{430}.

The purified forms of riboflavin and vitamin B_{12} would not be supplemented, due to the high cost and because both vitamins are synthesized by many bacteria present in most mixed systems. However, addition of excess sludge heat treatment liquor could conceivably supply such vitamins in restricted cases.

7.8 SUMMARY

The lack of understanding of nutritional requirements is often the cause of digestion problems. With proper nutrients, utilization rates can be increased by a factor as high as 15. Although nitrogen and phosphorus are the major nutrients required during anaerobic digestion, some of the other key nutrients are sulfides, iron, and nickel. From 3 to 23 mg/liter of sulfide is needed. The sulfide requirement is essentially equal to the nitrogen requirement. Although a required nutrient, sulfide is toxic in excess. Off-gas sulfide content of 6% is toxic. The addition of iron to digesters containing several thousand mg/liter of volatile acids can result in a rapid decrease to a few hundred mg/liter. Nickel additions of 1 mg/liter can also significantly increase utilization rates. With proper nutrients and process control, rates of volatile acid fermentation could be significantly increased.

The rate of activity of any class of microorganisms in the anaerobic digestion process appears to be a multiplicative function related to the concentration of all required nutrients. Absence of any one essential nutrient can stop the process. However, to our knowledge, the literature does not report the half-saturation concentration coefficient (K_S) for any of the inorganic nutrients — nitrogen, phosphorus, sulfide, iron, cobalt, nickel, molybdenum, tungsten, or selenium. Also, K_S is not reported for any trace organic nutrients either. This has a strong economic impact, because, with many candidate biomass feedstocks, nutrient supplementation is required, imposing an economic burden on the process. Excess supplementation of a nutrient is to be avoided, but the K_S curve must be defined to optimize the minimum concentration of any required nutrient. It would be helpful to develop a model based on chemical costs of each nutrient for process optimization concentration of any required nutrient. Such a model, if used to determine the rate and amount of methane production, would help to define the practical trade-off between sacrifice of process efficiency and cost of chemical supplementation, which may possibly arise with some of the more expensive trace organic stimulants.

Nitrogen and phosphorus requirements for CSTR operation can be satisfied with nutrient concentrations discussed; however, these levels may not be necessary in reactors designed for long retention of microorganisms. The total sulfide requirement for methanogenic cells may be equivalent to the nitrogen requirement for optimum CSTR performance. Therefore, key nutrients such as N, P, S, Fe, Ni, and Co should be studied to optimize minimum concentrations based on reactor type and operating conditions.

The complex chemical environment in an anaerobic digester makes it very difficult to predict the equilibrium concentration of heavy metals available in solution for microbial nutrition. The literature reports on the concentration of minerals added to the nutrient media. Even measurement of trace metals in solution is not a true measure of their availability to the microorganism. The metals may be chelated so strongly that the bacteria are unable to utilize them. Both trace metals and sulfide coexist in a digester in soluble and insoluble form, indicating the chemistry must be complex. The chemistry and chelation characteristics of anaerobic digestion environments, therefore, deserve special attention because they bear so strongly on the nutrient availability for the microorganism.

Nutritional requirements are often reported for batch growth systems. However, there are severe limitations in extrapolation of batch growth nutritional requirements to continuous culture systems typical of prototype biomass digesters. The continuous culture systems are very dynamic in nature; therefore, meaningful extrapolation of nutrient requirements derived from batch growth systems to continuous culture systems must be further examined.

The literature contains very little information on the nutrient requirements of hydrolytic, intermediate, and acetogenic bacteria. Relatively more information is reported on the nutrient requirements of the methanogens. Nutritional deficiency in the acetogens, hydrogen-utilizing methanogens, or acetate-utilizing methanogens could contribute to elevated volatile acids concentrations. Therefore, research is needed in these areas.

REFERENCES

1. Speece, R. E. and McCarty, P. L., 'Nutrient Requirements and Biological Solids Accumulation in Anaerobic Digestion,' *Adv. in. Wat. Poll. Res.* **2**:305–322, 1964.
2. Hoban, D. J. and van den Berg, L., 'Effect of Iron on Conversion of Acetic Acid to Methane During Methanogenic Fermentations,' *J. Appl. Bacteriol.* **47**:153–159, 1979.
3. Scherer, P. and Sahm, H., 'Effect of Trace Elements and Vitamins on the Growth of *Methanosarcina barkeri*,' *Acta Biotechnologica* **1**:57–65, 1981.

4. Speece, R. E., 'Environmental Requirements for Anaerobic Digestion of Biomass,' *Advances in Solar Energy, An Annual Review of Research and Development*, **2**:51–123, 1985.
5. Chynoweth, D. P., 'Microbial Conversion of Biomass to Methane,' Paper presented at the *Eighth Annual Energy Technology Conference and Exposition*, Washington, DC, March 1981.
6. Chynoweth, D. P., Jerger, D. E., Conrad, J. R., Razik, A., Srivastava, V. J., Ghosh, S., Henry, M. P. and Babu, S., 'Gasification of Land-Based Biomass,' Final Project Report for GRI, IGT Project No. 30564, Chicago, 1983.
7. Fannin, K. F., Srivastava, V., Chynoweth, D. P. and Bird, K. T., 'Effects of Interaction Between Biomass Composition and Reactor Design on Anaerobic Digestion Process Performance,' Paper presented at *Energy From Biomass and Wastes VII*. Chicago: Institute of Gas Technology, 1983.
8. Ghosh, S., Henry, M. P. and Klass, D. L., 'Bioconversion of Water Hyacinth-Coastal Bermuda Grass-MSW-Sludge Blends to Methane,' *Biotechnol. Bioeng. Symp.* **10**:163–187, 1980.
9. Zehnder, A. J. B., 'Ecology of Methane Formation,' Chapter 13 in R. Mitchell, Ed., *Microbiology of Water Pollution*. New York: Wiley Interscience, 1978.
10. Mah, R., Hungate, R. and Ohwaki, K., 'Acetate, a Key Intermediate in Methanogenesis,' pp. 97–106 in H. G. Schlegel, Ed., *Seminar on Microbial Energy Conversion*. Göttingen, Germany: E. Goltz KG, 1976.
11. Symons, J. M. and McKinney, 'The Biochemistry of Nitrogen in the Synthesis of Activated Sludge,' *Sew. and Indus. Wastes* **30**:874, 1958.
12. Hoover, S. R. and Porges, N., 'Assimilation of Dairy Wastes by Activated Sludge II. The Equation of Synthesis and Oxygen Utilization.' *Sew. and Ind. Wastes* **24**:306, 1952.
13. Frostell, B., 'Anaerobic Treatment in a Sludge Bed System With a Filter System,' *JWPCF* **53**:215, 1981.
14. Mountfort, D. O. and Asher, R. A., 'Effect of Inorganic Sulfide on the Growth and Metabolism of *M. barkeri*, Strain DM,' *Appl. Environ. Microbiol.* **37**:670–675, 1979.
15. Ronnow, P. H. and Gunnarsson, L. A. H., 'Sulfide Dependent Methane Production and Growth of Thermophilic Methanogenic Bacterium,' *Appl. Environ. Microbiol.* **42**:580–584, 1981.
16. Ronnow, P. H. and Gunnarsson, L. A. H., 'Response of Growth and Methane Production to Limiting Amounts of Sulfide and Ammonia in Two Thermophilic Methanogenic Bacteria,' *FEMS Microbiol. Lett.* **14**:311–315, 1982.
17. Gunnarsson, L. A. H. and Ronnow, P. H., 'Variation of the ATP Pool in Thermophilic Methanogenic Bacteria During Nitrogen or Sulfur Starvation,' *FEMS Microbiol. Lett.* **14**:317–320, 1982.
18. Patel, G. B., Khan, A. W. and Roth, L. A., 'Optimum Levels of Sulfate and Iron for the Cultivation of Pure Cultures of Methanogens in Synthetic Media,' *J. Appl. Bacteriol.* **45**:347–356, 1978.
19. Khan, A. W. and Trottier, T. M., 'Effects of Sulfur-Containing Compounds on Anaerobic Degradation of Cellulose to Methane by Mixed Cultures

Obtained from Sewage Sludge,' *Appl. Environ. Microbiol.* **35**:1027–1034, 1978.
20. Wellinger, A. and Wuhrmann, K., 'Influence of Sulfide Compounds on Metabolism of *Methanobacterium* Strain AZ,' *Arch. Microbiol.* **115**:13–17, 1977.
21. Bryant, M. P., Tseng, S. F., Robinson, I. M. and Joyner, A. E., 'Nutrient Requirements of Methanogenic Bacteria,' *Adv. Chem. Ser.* **105**:23–40, 1971.
22. Zehnder, A. J. B. and Wuhrmann, K., 'Physiology of a *Methanobacterium* Strain AZ,' *Arch. Microbiol.* **111**:199–205, 1977.
23. Wolfe, R. S., 'Microbial Formation of Methane,' pp. 107–146 in A. H. Rose and J. F. Wilkinson, Eds, *Adv. Microbiol. Physiol.*, Vol. 6. New York and London: Academic Press, 1971.
24. Lawrence, A. W. and McCarty, P. L., 'The Effects of Sulfides on Anaerobic Treatment,' *Int. J. Air and Wat. Poll.* **10**:207–221, 1966.
25. Wise, D., personal communication, 1982.
26. Lawrence, A. W. and McCarty, P. L., 'The Effects of Sulfide in Preventing Heavy Metal Toxicity in Anaerobic Treatment,' *J. Water Poll. Cont. Fed.* **37**:392–409, 1965.
27. Diekert, G., Jaenchen, R. and Thauer, R. K., 'Biosynthetic Evidence for Nickel Tetrapyrrole Structure of Factor F430 From *Methanobacterium thermoautotrophicum*,' *FEBS Letter* **119**:118–120, 1980.
28. Balch, W. E., Fox, G. E., Magrum, L. J., Woese, G. R. and Wolfe, R. S., 'Methanogens: Re-Evaluation of a Unique Biological Group,' *Microbiol. Rev.* **43**:260–296, 1979.
29. Diekert, G., Konheiser, U., Peichulla, K. and Thauer, R. K., 'Nickel Requirement and Factor F430 Content of Methanogenic Bacteria,' *J. Bacteriol.* **148**:459–464, 1981.
30. Mah, R. A., Smith, M. R. and Baresi, L., 'Studies on an Acetate-Fermenting Strain of *Methanosarcina*,' *Appl. Microbiol.* **16**:358–361, 1968.
31. Schonheit, P., Moll, J. and Thauer, R. K., 'Nickel, Cobalt and Molybdenum Requirement for Growth of *Methanobacterium thermoautotrophicum*,' *Arch. Microbiol.* **123**:105–107, 1979.
32. Jones, J. B. and Stadtman, T. C., '*Methanococcus vannielii*: Culture and Effects of Selenium and Tungsten on Growth,' *J. Bacteriol.* **130**:1404–1406, 1977.
33. Diekert, G., Klee, B. and Thauer, R., 'Nickel, a Component of Factor F430,' *Arch. Microbiol.* **124**:103–106, 1980.
34. Taylor, G. T. and Pirt, S. J., 'Nutrition and Factors Limiting the Growth of a Methanogenic Bacterium (*Methanobacterium thermoautotrophicum*),' *Arch. Microbiol.* **113**:17–22, 1977.
35. Winfrey, M. R. and Zeikus, J. G., 'Effect of Sulfate on Carbon and Electron Flow During Microbial Methanogenesis in Freshwater Sediments,' *Appl. and Environ. Microbiol.* **33**:275–281, 1977.
36. Speece, R. E., Parkin, G. F. and Gallagher, D., 'Nickel Stimulation of Anaerobic Digestion,' *Wat. Res.* **17**, 677–683, 1983.
37. Tanner, R. S., 'Novel Compounds From Methanogens: Characterization of Component B of the Methylreductase System and Mobile Factor,' PhD

Thesis, University of Illinois-Urbana, 1982.

38. Roberton, A. M. and Wolfe, R. S., 'ATP Requirement in Cell Extracts of *Methanobacterium Strain M.o.H.*,' *Biochem. Biophys. Acta*, 420–429, 1969.
39. Roberton, A. M. and Wolfe, R. S., 'Adenosine Triphosphate Pools in *Methanobacterium*,' *J. Bacteriol.* **102**:43–51, 1970.
40. Mountfort, D. O., 'Effect of Adenosine 5′-Monophosphate on Adenosine 5′-Triphosphate Activation of Methyl Coenzyme M Methylreductase in Cell Extracts of *Methanosarcina barkeri*,' *J. Bacteriol.* **143**:1039–1041, 1980.
41. Diekert, G., Weber, B. and Thauer, R. K., 'Nickel Dependence of Factor F430 Content in *Methanobacterium thermoautotrophicum*,' *Arch. Microbiol.* **227**:273–278, 1980.

8

Toxicity

RICHARD E. SPEECE
Environmental Studies Institute, Drexel University, Philadelphia, Pennsylvania, USA

8.1 INTRODUCTION

A common misconception about anaerobic digestion is that the process cannot tolerate toxic materials and that the biota 'die' when exposed to toxicants. This concept arose from municipal sludge digestion experiences when methane bacteria were inhibited by a toxicant. The result was that volatile acid production continued, exceeded the buffer capacity of the system, depressed the pH, and led to a 'sour' digester. This is a rather complex phenomenon in which the original toxicant may play only a minor 'triggering' role.

It is true that methanogens, like other organisms, can be inhibited by toxicants. Due to the relatively low fraction of substrate synthesized into cells by the methanogens and the associated prolonged generation times, the recovery period can be considerably extended if the toxicant is indeed bacteriocidal. However, studies on toxicity recovery of only the methanogens indicate that most toxicants exhibit a bacteriostatic or reversible effect on the methanogens at the lower concentration ranges normally encountered in industrial wastewaters.[1]

The effect of molecular structure (using 52 petrochemicals) was assayed with unacclimated acetate-enriched methanogens.[2] The following structural characteristics influenced the selection of the petrochemicals that were assayed: double bonds, position of functional group, length of carbon chain, branching, oxidation state, type of functional groups, number of identical functional groups, and configuration. The results showed a definite correlation between molecular structure and toxicity to unacclimated methanogens. Chloro substitution, aldehydes, double bonds, and benzene rings exhibited toxicity. Addition of hydroxyl groups and increased carbon chain length decreased toxicity. Acclimation to toxic compounds substantially reduced their toxicity.

8.2 ACCLIMATION AND METABOLISM

A correlation between the toxicity characteristics of an organic compound and an unacclimated, acetate-enriched methanogenic culture was not shown in a study with petrochemical wastewater components.[3] Toxicity was interpreted as meaning no metabolism and acclimation was interpreted as meaning the substance was actually metabolized. For example, aldehydes showed toxic inhibition to an unacclimated, acetate-enriched methanogenic culture, whereas ketones exhibited no toxicity. However, upon acclimation, aldehydes actually degraded more readily than ketones.

Healy and Young[4] observed that the time required for aromatic compounds to be degraded via methane fermentation depends on substrate structure and the source of inoculum. These authors also studied cross-acclimation of potential aromatics formed during the heat treatment of lignin. They suggested that a microbial population acclimated to a particular aromatic substrate can be simultaneously acclimated to other aromatic substrates with similar arrangements of side groups on the aromatic ring. Weng and Jeris[5] reported that glutamic

TABLE 8.1
Petrochemicals Metabolized by Enriched Methane Cultures[2]

Acetaldehyde	Formaldehyde	Phthalic acid
Acetone	Formic acid	Propanal
Adipic acid	Fumaric acid	Propanol
1-Amino-2-propanol	Glutaric acid	2-Propanol
4-Amino-butyric acid	Glycerol	Propionic acid
Benzoic acid	Hexanoic acid	Propylene glycol
Butanol	Hydroquinone	Resorcinol
Butyraldehyde	Isobutyric acid	sec-Butanol
Butyric acid	Maleic acid	sec-Butylamine
Catechol	Methanol	Sorbic acid
Crotonaldehyde	Methyl acetate	Succinic acid
Crotonic acid	Methyl ethyl ketone	tert-Butanol
Ethyl acetate	Nitrobenzene	Valeric acid
Ethyl acrylate	Phenol	Vinyl acetate

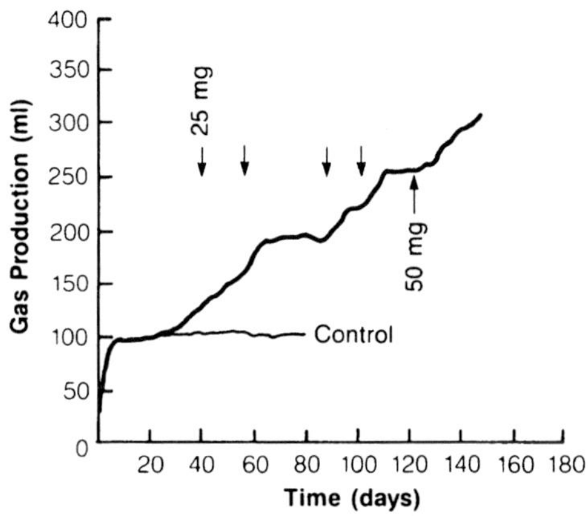

FIG. 8.1. Cross-acclimation of acetate culture to phenol.[2]

acid-acclimated cultures could utilize acetic, propionic, mesaconic, keto-glutaric, and pyruvic acids, converting them to methane and carbon dioxide. Chou *et al.*[3] assayed the biodegradability of a wide spectrum of petrochemicals by methanogenic acetate enrichment cultures. Table 8.1 shows the petrochemicals that were eventually metabolized by the acetate-enriched, methane-producing cultures.

Figure 8.1 shows the cross-acclimation characteristics of an acetate-enriched methane-producing culture to phenol. The control received only acetate. After approximately 30 days the culture started to metabolize the phenol, even though this compound is considered a disinfectant at high concentrations. The same is true with formaldehyde. However, at dilute concentrations, both toxicants were readily metabolized. Speece[1] fed an anaerobic filter a wastewater that contained 1000 mg/liter phenol and 5 mg/liter cyanide, and the system converted the phenol to methane at a process efficiency in excess of 90% at a hydraulic retention time of 12 hours.

The structural characteristics of compounds that affect cross-acclimation and metabolism were recognized as follows:

(1) Nontoxic aliphatic compounds containing carboxyl, ester, or hydroxyl groups were all readily metabolized by acetate-enriched cultures. Butyric acid, valeric acid, propanol, propylene glycol, ethyl acetate, and vinyl acetate were all metabolized within 4 days.

(2) Compounds with carbonyl functional groups or double bonds were metabolized within a short period of time (butyraldehyde and sorbic acid required 7–10 days to acclimate although these compounds exhibited toxicity to unacclimated acetate enriched cultures).
(3) Compounds with amino functional groups showed difficulty in being metabolized. Aniline, 4-amino-butyric acid, sec-butylamine, and 1-amino-2-propanol degraded very slowly.
(4) Dicarboxylic acids required a longer period of time to metabolize than monocarboxylic acids of the same carbon chain length.
(5) The position of the functional group affected the lag period for metabolism; for example, primary butanol required 4 days for metabolism, secondary butyl alcohol required 14 days and tertiary butyl alcohol was not metabolized in suspended growth systems, but eventually was metabolized in an anaerobic filter.

8.3 MICROBIAL KINETICS AND TOXICITY

Industrial wastewaters commonly contain chronic or transient levels of toxicity. Consequently, a wastewater treatment system receiving industrial wastes would rarely operate at true steady-state. For instance, under ideal conditions, the microbial regeneration time of methanogens metabolizing acetate would be approximately 2–4·2 days.[6,7] However, when a toxicant is present in the wastewater, the specific utilization of acetate may be reduced (noncompetitive inhibition) or the half-velocity constant of the substrate, K_S, may be increased (competitive inhibition). The minimum biological SRT, $\theta_{c\,min}$, is defined as:

$$\theta_{c\,min} = (Y_G q - k_d)^{-1} \tag{8.1}$$

where Y_G is the growth yield constant, q is the rate of substrate utilization, and k_d is the endogenous respiration rate. Therefore, anything that decreases the product of the bacterial yield, Y_G, and q will increase the generation time. Since a toxicant inhibits substrate utilization, it increases the microbial generation time.

The 'dynamic' nature of $\theta_{c\,min}$ with toxicant exposure is demonstrated in Fig. 8.2. After addition of the toxicant, $\theta_{c\,min}$ increases with concomitant decreases in methane production and process efficiency. Although presented as steady-state response for each condition, the 'elastic' nature

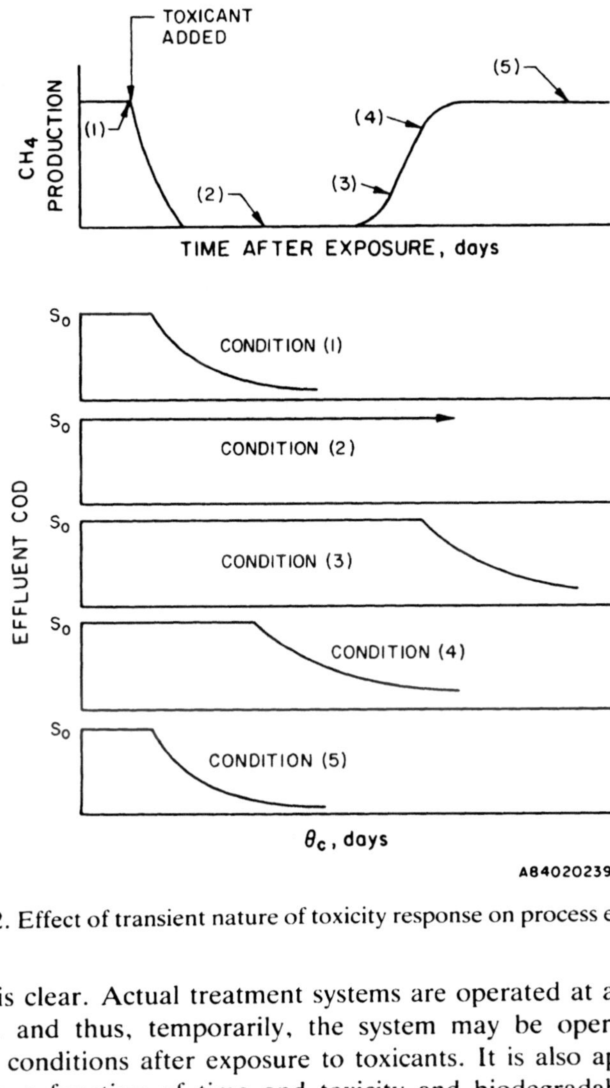

FIG. 8.2. Effect of transient nature of toxicity response on process efficiency.[2]

of $\theta_{c\,min}$ is clear. Actual treatment systems are operated at a reasonably fixed θ_c, and thus, temporarily, the system may be operating under washout conditions after exposure to toxicants. It is also apparent that failure is a function of time and toxicity and biodegradability of the toxicant. A sufficient biological safety factor (defined as $\theta_c/\theta_{c\,min}$) in the form of a proper design, θ_c, guards against the above-described failure and allows for acclimation. For example, Chou *et al.*[3] could not achieve metabolism of acrylic acid in a suspended-growth, acetate-enriched culture with a 25-day hydraulic and solids retention time, because the microbial cells washed out before acclimation occurred. However, acrylic acid was readily metabolized in an anaerobic filter having a solids retention time in excess of 100 days. Thus the generation time of the culture was apparently in excess of 25 days during the acrylic acid acclimation phase. The need for prolonged solids and microorganism retention times in methane fermentation processes treating industrial wastewaters is not sufficiently recognized.

8.4 TOXICITY IN ANAEROBIC DIGESTION STUDIES

Suspended-growth systems of acetate-enriched methanogens that have received a wide variety of toxicants all show a characteristic recovery pattern when a neutral pH is maintained and the acetate concentration is restored daily to the range of 1000–2000 mg/liter. Figure 8.3 shows the response of a suspended-growth system to chloroform. Gas production rate was inhibited and for a time was proportional to the toxicant concentration. Even after prolonged periods of nil gas production, eventual recovery was very rapid and complete. This indicated that biomass viability was maintained and that the toxicity was bacteriostatic, not bacteriocidal.[8] A similar response was also shown with cyanide, ammonium and oxygen.

The reversibility of toxic inhibition was confirmed in two ways.[8] First, the feed to plug-flow anaerobic filters was spiked with various toxicants. The plug-flow anaerobic filters resulted in almost complete displacement of the feed and thus of the toxicant spikes in less than 2 days. Figure 8.4 indicates that when 2000 mg/liter of sulfide was added to the feed of a filter for a day, gas production almost ceased. However, after 2 days gas production was back to normal. The second method involved experiments with suspended growth systems. Very high concentrations of formaldehyde (300, 600, 1800 mg/liter) were injected into duplicate systems. One of the duplicates was centrifuged after exposure for 1 hour. The supernatant was then discarded and replaced with an unadulterated nutrient salt solution. The other duplicate served as a control with no removal of the soluble fraction. Figure 8.5 shows that gas production in the systems exposed for 1 hour to 300 or 600 mg/liter of formaldehyde exhibited only a slight, temporary reduction in gas production. The system exposed to 1800 mg/liter of formaldehyde for 1 hour took about 10 days to recover to 75% of the pretoxicant gas production rate. The gas production ceased in the controls.

High sulfate concentrations are common in some biomass feedstocks

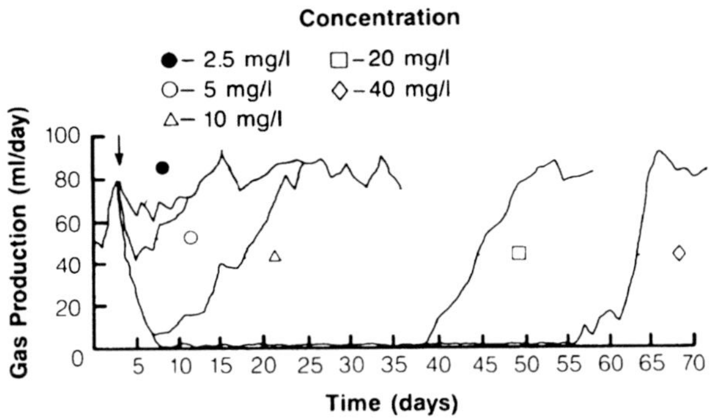

FIG. 8.3. SRT response to toxicant chloroform (50 days SRT, CSTR, 35 °C).[1]

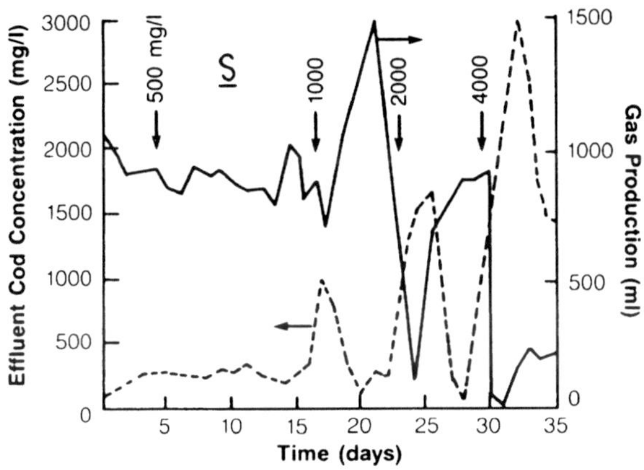

FIG. 8.4. Response of anaerobic filter to toxicant (sulfide).[1]

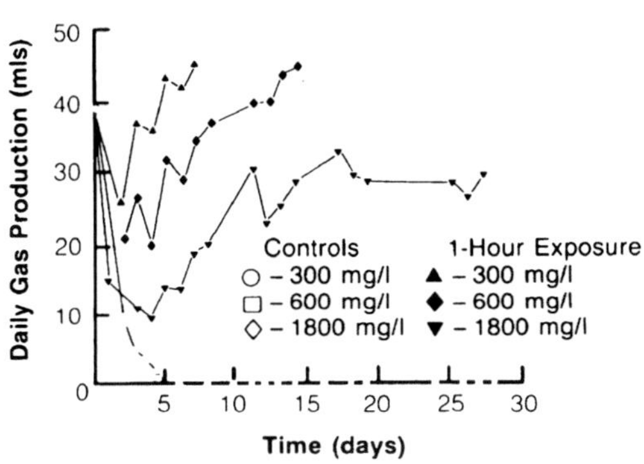

FIG. 8.5. Effect of exposure time on toxicity response (1 h exposure to formaldehyde).[1]

and industrial wastewaters. Sulfate reduction with hydrogen is energetically favored over methane production. Yet it has been observed that sulfate reduction is not complete in some pilot plant studies of anaerobic digestion. Since excessive hydrogen sulfide concentrations can be toxic, it would be advantageous to be able to inhibit hydrogen sulfide production without inhibiting methane production. Identification of such a selective inhibitor for hydrogen sulfide would be of academic importance and, depending on the concentration required and unit cost, it could be of practical value in field scale digesters.

Pesticides are used in agriculture to produce greater crop yields, crops of higher quality, and to reduce the input of labor and energy into crop production. If these practices are applied to biomass energy forms, toxicity and inhibition of microorganisms due to pesticide residues would affect digestion process kinetics. A recent review discusses the microbial biodegradability of pesticides along with the toxicity of these compounds to aerobic and anaerobic microorganisms.[9] Softwood resin extracts consist of resin acids, fat, and terpenes and also contain a variety of phenolic compounds such as flavonoids, ligands, and stilbenes. Some pine species have been reported to contain up to 14% dry wt extractives.[10] Water-soluble tannins are also present in softwoods. Some of these compounds are referred to as 'pathological' resins as they protect the tree against biological damage. Hardwood resins typically consist of fats, waxes, and sterols, but species such as *Eucalyptus* sp. have been reported to be rich in isoprenoid-containing compounds.[11] These extractives may be inhibitory to the digestion process and should be studied to determine their effects on biomass degradability.

Some form of pretreatment of biomass feedstocks is desirable to increase yields. Acid or base hydrolysis has shown promise. However, in the process, byproducts can be formed, such as furfural, hydroxymethyl furfural, formic acid, and levulinic acid, which exhibit inhibition. Potential acclimation to such inhibition and perhaps eventual metabolism may be possible, but process kinetics may be affected.

Kugelman and Chin[12] made the following observations:

> A review of the literature on anaerobic waste treatment indicates considerable variation in the toxicity reported for most substances. In some cases, the range is over more than one order of magnitude. The major reason for these variations is the complexity of the phenomenon of toxicity. Unfortunately, in most studies of toxicity in anaerobic waste treatment, little or no attention is paid to the effects

of antagonism, synergism, acclimation, and complexing reactions. Indeed, in very few studies are the chemical characteristics of the control defined adequately. The main reasons for these experimental deficiencies are:

Much published toxicity data for anaerobic waste treatment are the product of studies in which toxicity was not to be studied! Rather, toxicity developed during a study of some other phenomenon, and the toxicity data were reported, sometimes as an afterthought, with the other data.

Another main source of toxicity data is from reports on failure of and attempts to revive field digestion systems. These reports were primarily filed by treatment plant operators who were not equipped to run the sophisticated chemical analyses required to categorize adequately the environment in the digester. Neither, for that matter, did they have the time, since coping with a field scale 'sour' digester is a formidable task.

Most studies have been conducted in full strength or diluted sewerage sludge, which is an incredibly complex substance whose characteristics vary with geographical location. This procedure has been necessary because it is difficult to develop a chemically defined medium in which methane bacteria will flourish. These organisms will thrive in sewage sludge.

Although there are good reasons why much of the data on toxicity in anaerobic systems are suspect, such data have not and cannot be interpreted properly. At best, most of these data are useless; at worst, they are misleading.

8.5 SUMMARY

Several general observations on toxicity in anaerobic digestion indicate that methanogens show the same general response pattern to a wide variety of toxicants. However, the toxicity in many cases appears to be reversible. Acclimation to the toxicity is demonstrated with methanogens for some toxicants. These observations demonstrate that methanogens can tolerate toxicity and, therefore, provide considerable optimism on the potential of anaerobic digestion.

With most soluble feedstocks for bioconversion to methane, the rate-limiting step is hydrolysis, not methanogenesis. Yet there is even less information on the effects of various toxicants on the hydrolysis, acid fermentation, or acetogenic steps of anaerobic digestion than for the methanogenesis step. If the digestion process is to be optimized, the rate-limiting step must be enhanced.

The mechanism by which a toxic substance inhibits bacterial activity is related to the toxicant. In aerobic bacterial metabolism, general biochemical pathways are defined and the mechanism of toxic inhibition has been studied more extensively. Toxicants have been grouped into classes, depending on the mechanism of toxicity. In anaerobic digestion, the biochemical pathways are not all defined. For instance, the pathway for methane production is not completely known. Therefore, it is more difficult to classify toxicants in anaerobic digestion.

There is no conclusive correlation between the toxicity characteristics of an organic compound to an unacclimated acetate-enriched methanogenic culture and the eventual metabolism of that same organic by the acclimated culture. Perhaps this is also true of the entire consortium of bacteria in anaerobic digestion. A clear distinction needs to be drawn between unacclimated and acclimated toxicity. Yet there is essentially no information on the correlation between the exposure time of a toxicant to microorganisms and the degree of resulting acclimation. Neither is there a comparable data base showing the effect of toxicant concentration on acclimation potential. In order for the realization of the full potential of the consortium of bacteria in anaerobic digestion to accommodate chronic or sporadic toxicity, the relation between type of toxicant, concentration, and duration of exposure must be known.

The acclimation phenomenon is observed, but not understood. Perhaps if it were better understood it could be more effectively exploited. Of the wide variety of toxicants assayed for methanogenic response, it would seem that the specific site of damage within the cell would be dependent on the toxicant. Yet the gross observation of methanogens based on methane production for the toxicants assayed indicated that the cell is eventually able to repair the damage wherever it is incurred. Full restoration of the original rate of methane production has been observed after toxicant exposure even though the toxicant concentration is unchanged. There appears to be great potential for methanogen acclimation potential to toxicity.

Since biomass digestion requires a complete consortium of bacteria, it is relevant to study the effect of a given toxicant and its potential acclimation on the overall process as well as to individual steps within the process. A toxicant can inhibit the rate-limiting step and/or change the step that is rate limiting. Both manifestations of toxicity can severely affect the overall process.

Propionic acid and, to a lesser extent, some of the higher volatile acids have been reported to accumulate in malfunctioning digestion systems and also during start-up. Acetogenesis is such a key step that specific effort should be concentrated here. Certain acetogens can only be cultured in combination with hydrogen-utilizing methanogens and it is difficult to determine analytically hydrogen at the low concentrations required for acetogenesis to proceed. Therefore, acetogenic toxicity assay will be more difficult to evaluate because of this interdependency, but it should be investigated nevertheless because of its key role.

The reversibility of toxicity strongly affects the choice of digester configuration. Of the limited observations of this phenomenon in methanogenic assays, it appears that plug flow results in less overall reduction in process efficiency than CSTR. This needs to be clarified for the hydrolytic, acid fermenting, and acetogenic steps.

There is the need for a more comprehensive viability assay for the microorganisms responsible for all of the steps of anaerobic digestion. (Methane production is not sufficiently comprehensive for assaying the viability of methanogens.) Cessation of methane production has been shown *not* to correlate with the viability of methanogens. ATP and coenzyme F_{420} concentrations also have been shown *not* to correlate with cell viability. Toxicity response would be greatly enhanced by a comprehensive viability assay for microorganisms in each stage of anaerobic digestion.

Antagonism and synergism have been clearly demonstrated with iron toxicity. Specific treatments have been reported for remedial treatment of toxicity due to anionic detergents, cyanide, and chloroform. There is a need to expand this list of remedial treatments for additional toxicants.

REFERENCES

1. Speece, R. E., 'Environmental Requirements for Anaerobic Digestion of Biomass,' *Advances in Solar Energy, An Annual Review of Research and Development.* **2**:51–123, 1985.
2. Chou, W. L., Speece, R. E., Siddiqi, R. H. and McKeon, K., 'The Effect of Petrochemical Structure on Methane Fermentation Toxicity,' *Progr. Water Technol.* **10**:545–550, 1978.
3. Chou, W. L., Speece, R. E. and Siddiqi, R. H., 'Acclimation and Degradation of Petrochemical Wastewater Components by Methane Fermentation,' *Biotech. and Bioeng. Symp.* **8**:391–414, 1978.
4. Healy, J. B., Jr and Young, L. Y., 'Anaerobic Biodegradation of Eleven Aromatic Compounds to Methane,' *Appl. Environ. Microbiol.* **38**:84–89, 1979.

5. Weng, C. and Jeris, J. S., 'Biochemical Mechanisms in the Methane Fermentation of Glutamic and Oleic Acids,' *Wat. Res.* **10**:9–18, 1976.
6. Lawrence, A. W. and McCarty, P. L., 'Kinetics of Methane Fermentation in Anaerobic Treatment,' *J. Water Poll. Cont.* **41**:R1–R17, 1969.
7. Huser, B. A., Wuhrmann, K. and Zehnder, A. J. B., '*Methanothrix sohngenii* Non-Hydrogen Oxidizing Methanobacterium,' *Arch. Microbiol.* **132**:1–9, 1982.
8. Yang, J., Speece, R. E., Parkin, G. F., Gossett, J. and Kocher, W., 'The Response of Methane Fermentation to Cyanide and Chloroform,' *Progr. Water Technol.* **12**:977–989, 1980.
9. Hill, I. R. and Wright, S., *Pesticide Microbiology*. New York: Academic Press, 1978.
10. Sjostrom, E., *Wood Chemistry: Fundamentals and Applications*. New York: Academic Press, 1981.
11. Sachs, R. M., Gilpin, D. W. and Mock, T., 'Yields of Short Rotation *Eucalyptus grandis* in High Density Plantings,' in Symposium Papers, *Energy From Biomass and Wastes IV*. Chicago: Institute of Gas Technology, 1981.
12. Kugelman, I. J. and Chin, K. K., 'Toxicity, Synergism and Antagonism in Anaerobic Waste Treatment Processes,' *Anaerobic Biological Treatment Processes*, Advances in Chemistry Series **105**:55–90, 1971.

9
Reactor Designs

KIRBY F. FANNIN and RICHARD BILJETINA
Institute of Gas Technology, Chicago, Illinois, USA

9.1 INTRODUCTION

To develop economically attractive commercial technology for anaerobic digestion of biomass feedstocks, it is necessary to match the physical and chemical properties of the feedstock with design and performance considerations. Selection of the proper digester design for a particular feedstock is a key element in the development of an appropriate process. Therefore, an understanding of the operating principles of the major digester types may help in selecting a processing scheme and possibly reducing the extent of the development effort.

The relative importance of such performance parameters to high methane yields and production rates is somewhat dependent on the relative contribution of feedstock and reactor costs to the final gas cost. When the feedstock costs are high, high methane yields can have a greater effect on reducing gas costs than high methane production rates. When feedstock costs are low, on the other hand, higher methane production rates are more important for lower gas costs. High methane yields can be achieved through long solids and microorganism retention times (SRTs and MRTs), whereas high methane production rates require high organic loading rates and short hydraulic retention times (HRTs).

The retention of biomass solids, microorganisms, and biomass liquids is determined by reactor design and operating techniques and by feedstock composition. Long SRTs can be attained by reducing the loading rate or by retaining the solids but removing the liquid. This latter procedure permits both long SRTs and short HRTs. The solids content of different biomass feeds determines the HRT and the feed heating requirements. Short HRTs can be maintained while promoting longer SRTs and MRTs by using several unconventional reactor designs.

9.2 RETENTION TIME

9.2.1 Hydraulic

The HRT in anaerobic digesters is determined by calculating the number of days required for displacement of the fluid volume of the culture. At a given organic loading rate, the HRT is lower when using high-water-content feeds than when using those containing less water. When undiluted feed is used, the HRT in a digester operated at a given organic loading rate is a function of the water content of the biomass. As the HRT is reduced by either using a feed containing lower solids or by increasing the loading rate, or both, increasingly large numbers of bacteria are washed out of the system. Unless some measure is taken to retain these solids and organisms, decreased conversion efficiency, lower methane yields, and greater digester instability are likely to occur. The slower-growing methanogenic bacteria will wash out of the reactor more rapidly than the faster-growing, acid-producing bacteria, causing a microbiological imbalance, accumulation of high levels of volatile acids, and eventual destabilization of the system.

Discussions of conventional digestion have emphasized the importance of HRT because it was the same as the SRT and determined the reactor size. In reactors that promote retention of solids and microorganisms, however, HRT is not a very useful parameter for defining the digester operation. In terms of large-scale biomass digester operation, the critical problem is to determine the amount of the reactive component feed that can be added to the digester per unit of time, not how much water is retained per day. As discussed below, the performance of the digester is more dependent upon SRT than HRT.

9.2.2 Solids

The SRT in a well-mixed system is the same as the HRT. In a completely mixed continuously stirred tank reactor (CSTR), for example, the time required for replacement of the digester contents is determined by the following expression:

$$C_t/C_0 = e^{-t/\theta} \tag{9.1}$$

where C_t = concentration at time t, C_0 = initial concentration, t = time (days), and θ = retention time (days). Using such an expression, it can be determined that three retention periods are needed to replace 95% of the digester contents.

In digesters that are not completely mixed, such as upflow solids reactors (USR), the SRT must be determined empirically. Provided the concentration and density of the suspended solids within the reactor and in the effluent can be determined,[1] the SRT in such reactors can be calculated as follows:

$$SRT = [(TSS_r)(RV \times D_r)]/[(TSS_e)(EV \times D_e)] \quad (9.2)$$

where SRT = solids retention time (days), TSS_r = average percent TSS concentration in reactor, TSS_e = percent TSS concentration in reactor effluent, RV = reactor volume, EV = effluent volume per day, D_r = density of solids in reactor, and D_e = density of solids in reactor effluent.

The effect of increased SRT on the degradation of particulate volatile solids using raw sludge is illustrated in Fig. 9.1.[2] These data indicate that volatile solids degradation followed a first-order decay model, assuming 20% nonbiodegradable solids at an SRT of up to 25 days. At longer SRTs, using a kelp feedstock, Fannin *et al.*[3] showed similar performance, as illustrated in Fig. 9.2. As the SRT exceeded 25 days, however, the data no longer followed the Michaelis-Menten kinetic model. As illustrated in

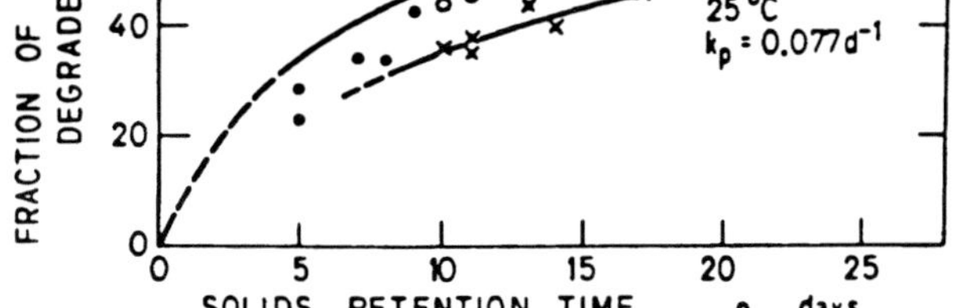

FIG. 9.1. Degradation of particulate volatile solids in raw sludge as a function of solids retention time. Continuous lines are predicted by a first-order decay model with an assumption of 20% nonbiodegradable solids. Closed circles indicate hydraulic retention time (HRT), θ_h, of 5 days at 35 °C; open circles indicate HRT of 10 days at 35 °C; crosses indicate HRT of 10 days at 25 °C.

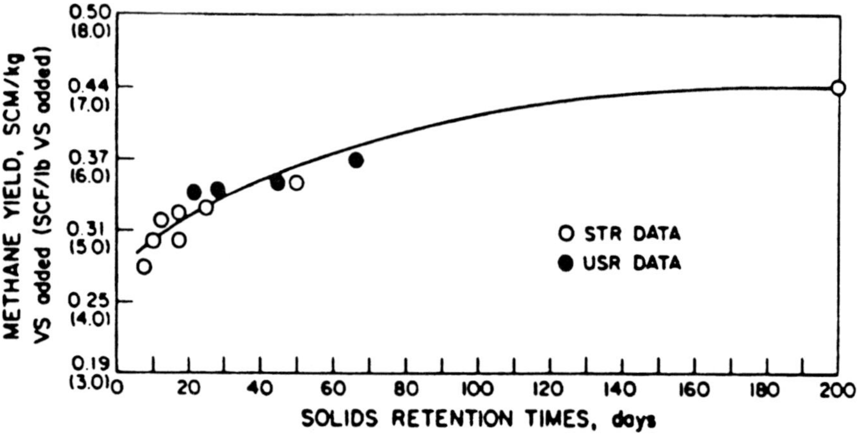

FIG. 9.2. Effect of solids retention time on methane yield of sea kelp.

Fig. 9.3, the concentration of biodegradable substrate in kelp CSTR digesters remained nearly constant. These data suggest that, at long retention times, the substrate was degraded and became unavailable for further conversion.

Digesters operated at long SRTs have lower bacterial cell synthesis requirements and, consequently, lower energy and nutrient requirements.[3] The ammonia nitrogen concentrations are higher in digesters operated at longer SRTs. Kelp digesters operated at 35 °C and SRTs ranging from 7 to 200 days, for example, demonstrated that ammonia nitrogen concentration increases from about 110–220 mg/liter to nearly 800 mg/liter as the SRT is increased (Fig. 9.4).

Using an anaerobic fluidized-bed reactor to treat moderate strength lactic casein whey permeate, nutrient requirements were lower than for CSTR systems, eliminating the need for supplemental nitrogen and phosphorus.[4] Nutrient requirements are lower for reactors that promote longer SRTs than HRTs. This is due either to decreased bacterial cell synthesis requirements and/or increased degradation in nutrient-containing compounds.

These observations from several digesters operated at different SRTs support the findings by Speece and McCarty[5] that the nitrogen requirements for anaerobic digestion can be reduced by increasing the SRT. Because of lower nitrogen requirements, a higher carbon-to-nitrogen ratio may be required in feeds that are digested in reactors that promote

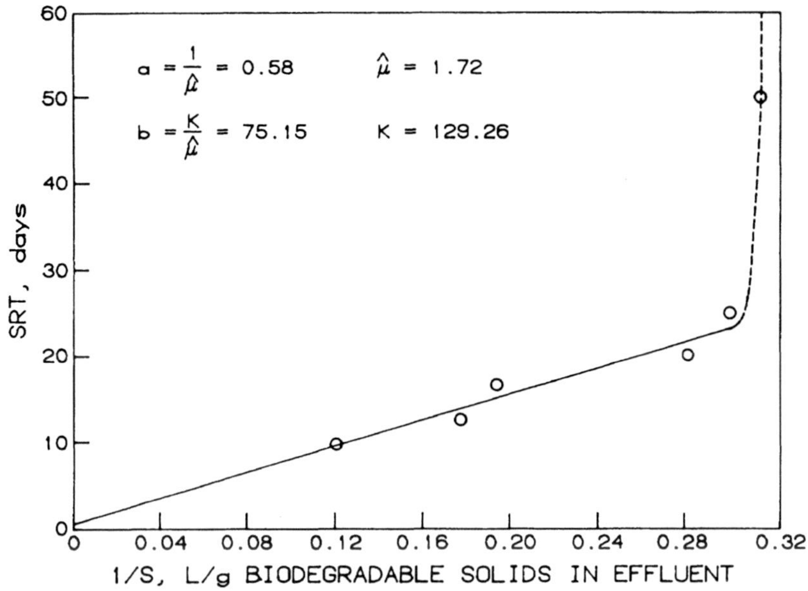

FIG. 9.3. Biodegradable effluent solids concentration at several SRTs in CSTRs operated on kelp at 35 °C.

longer SRTs (discussed in Chapter 7). The ammonia-enriched effluent from such reactors will also have a higher value as a potential fertilizer.

9.2.3 Microorganisms

One important method for stabilization of the biomass digester at very short HRTs is to promote the retention of microorganisms in the system. Increased microorganism retention prevents development of imbalance by preventing washout of slow-growing methanogenic bacteria. It also reduces the nutrient requirement for cell synthesis. Microorganism retention can be accomplished by employing attached film or some other microorganism-retaining reactor.

9.2.4 Effect of Biomass Solids Concentration

The choice of reactor design and mode of operation are highly dependent upon the physical and chemical characteristics of the biomass feeds, especially the concentration of particulate-associated solids. Based on laboratory-scale studies, biomass feedstocks were classified by Fannin *et*

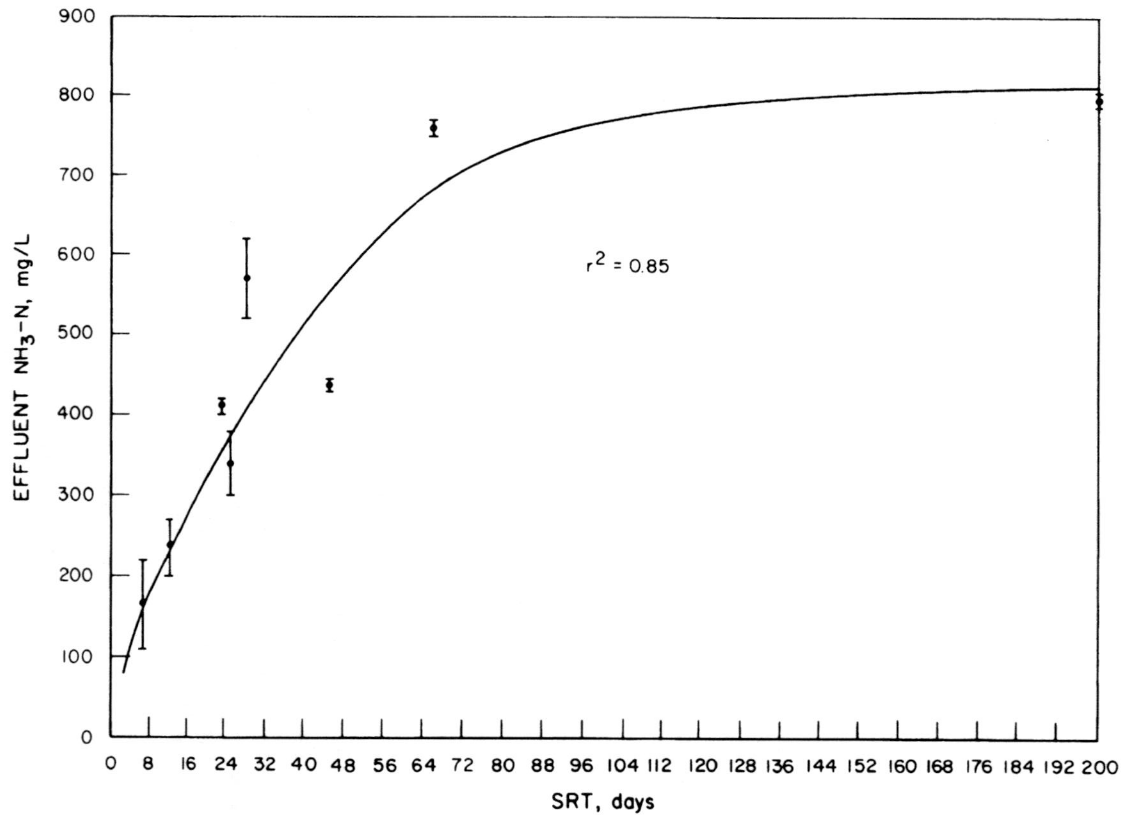

FIG. 9.4. Relationship between SRT and ammonia nitrogen in mesophilic kelp digester effluent.

TABLE 9.1
Types of Biomass Used in Anaerobic Digestion

*Biomass type**	*Example*
Soluble (<2% solids)	Water hyacinth juice
Low solids (2–5% solids)	Water hyacinth
Medium solids (6–15% solids)	Kelp
High solids–low lignin (>15% solids)	Sorghum, water hyacinth cake
High solids–high lignin (>15% solids)	Hybrid poplar

*Particulate-associated solids.

al.[3] into general categories, shown in Table 9.1, determined by their relative solids content.

The solids concentration of the biomass feeds has a direct impact on the HRT in a reactor system. This concept is illustrated in Fig. 9.5 for several biomass feeds with different solids concentrations. At a loading rate of 0·14 kg VS/m^3-day, for example, the HRT ranges from about 2·3 to 103·0 days for water hyacinth juice and hybrid poplar (diluted to 50%), respectively.

Increases in feed solids concentration at a given HRT increase the loading rate. At some upper solids concentration limit and at low SRTs, additional biomass substrate will remain unreacted because of either mass transport limitations or saturation of the organisms' enzyme systems. Recent experimental data[6] indicated that methane production became limited at 32% solids concentrations, while hydrolysis continued until the solids concentration reached approximately 60%. Batch conversion efficiencies of 90% of the biodegradable matter were achieved at an initial solids concentration of 28% in 200 days.

9.3 REACTOR CONFIGURATION

Anaerobic digestion of biomass can be performed in a variety of reactor designs that promote different HRTs, SRTs, and MRTs (microorganism retention times) when operated under similar loading conditions. These reactors are operated on different principles and their relative performance is affected by the biomass feedstock composition. Since biomass feeds vary widely with regard to their chemical composition and

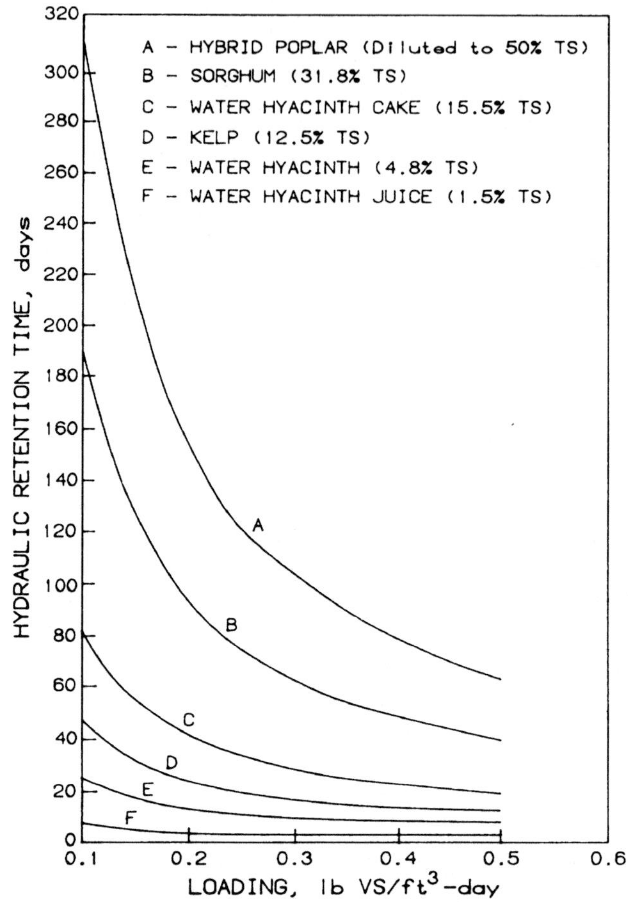

FIG. 9.5. Hydraulic retention time for several biomass feeds at specific loading rates.

physical properties, different species require different reactor designs for optimum performance. Biomass conversion reactors may be classified on the basis of the relative period of biomass liquid, biomass solids, and microorganism retention. Three general biomass reactor classifications are listed in Table 9.2.

9.3.1 Reactor Category A

9.3.1.1 Batch-Fed Reactors

In batch-fed digesters (Fig. 9.6) the total feed volume is added to the reactor where it remains until digestion is completed. Because the total

TABLE 9.2
Biomass Reactor Classification

Category	Retention characteristics	Reactor examples
A	Microorganisms = Solids = Liquid	Continuously stirred tank Plug flow (ideal) Batch fed
B	(Microorganisms and solids) > Liquid	Continuously stirred tank with solids recycle (contact process) Upflow solids Upflow sludge blanket Baffle flow
C	Microorganisms > (Solids and liquid)	Anaerobic filter Fluidized and expanded bed

feed is retained, the SRT, HRT, and MRT are the same. This represents a noncontinuous process that is particularly suited for seasonally produced biomass feeds and for feeds with a very high solids content. While these reactors may be mixed, larger-scale reactors, especially those using high solids content feeds, are frequently nonmixed.

These systems are relatively inexpensive to construct and operate compared to CSTRs. Open pits, such as quarries, can be utilized or smaller covered pits can be utilized.[7] High solids digesters require an initial mixture of inoculum, organic substrate, nutrients, and buffer to insure successful anaerobic digestion in a batch-fed system.[7] Nutrients and buffering capacity may also be added to the batch digester through recycled leachate.

A sanitary landfill could be considered to be an example of a large-scale batch reactor. While conventional landfills are typically not optimized for methane production, controlled landfill studies are under way[8] and novel approaches to improving the production and recovery of methane have been developed.[9] Another high solids digestion concept was referred to as 'dry fermentation'.[10] Using this system, batch conversion efficiencies of 90% of the biodegradable matter were reported at an initial solids concentration of 28% in 200 days. The concept shows potential for application to agricultural products using in-place production and collection techniques. It minimizes the addition of water and, therefore, energy consumption; it is, in principle, simple in both design and operation.[11]

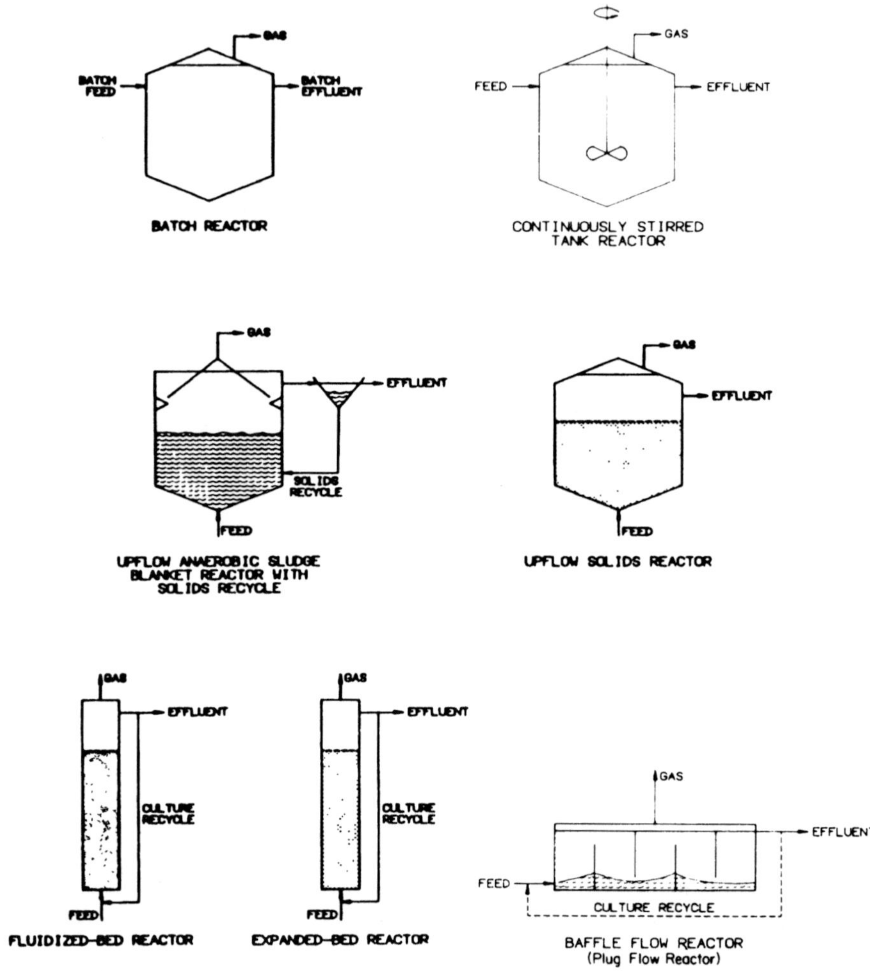

FIG. 9.6. Biological gasification process schemes.

9.3.1.2 Continuously Stirred Tank Reactors (CSTRs)

These reactors, illustrated in Fig. 9.6, are widely used and considered conventional for sewage sludge digestion. The design incorporates stirring or agitation of the contents to achieve good mixing using pumps, mechanical stirrers, gas recycle, or other options. The heated and mixed

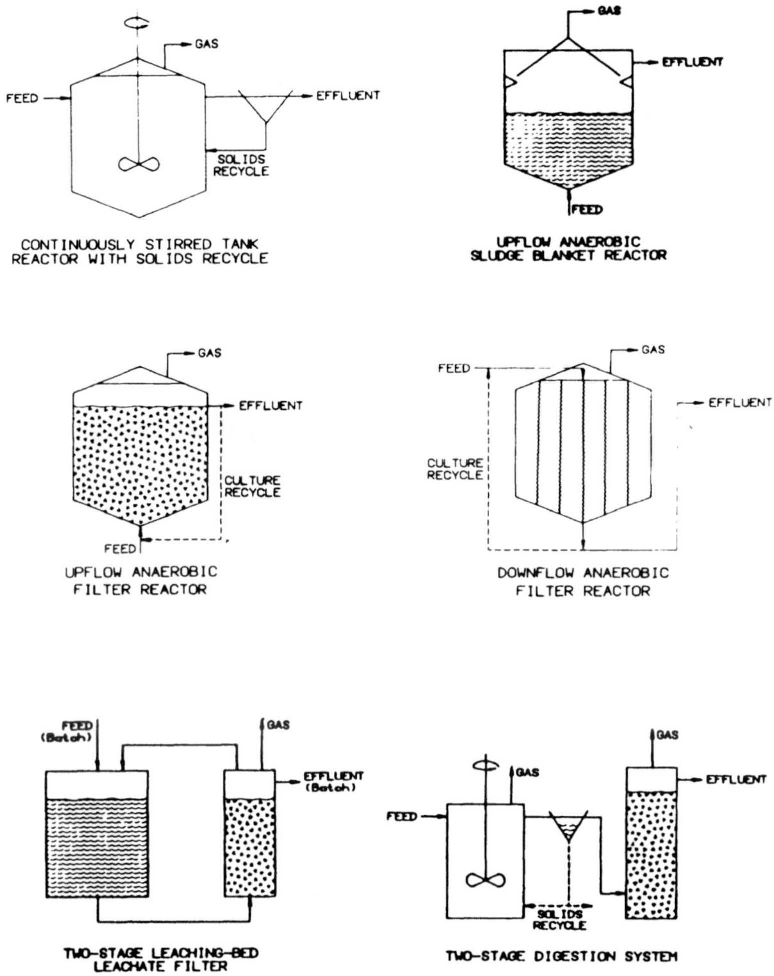

FIG. 9.6—*contd.*

reactor concept ensures that anaerobic digestion takes place at a stable optimum temperature. Operated as a once-through completely mixed reactor, the HRT in CSTRs is equal to the SRT and MRT. Large reactors may be required to achieve minimum required SRTs, which often exceed 20 or more days. Because of mixing, the effluent contains the same concentration of bacteria as the digester itself. Consequently, the number of bacteria lost upon effluent wasting increases at higher hydraulic loading rates. If the rate of bacterial washout exceeds their growth rate, the number of bacteria in the system will decrease both conversion efficiency and digester stability.

There are both advantages and disadvantages to the stirred tank reactor concept.[6] Some advantages reported include:

(1) Ability to process feeds with high levels of suspended solids.
(2) Enhanced contact of microbiological species with the substrate.
(3) Uniform distribution of the substrate throughout the tank.
(4) Prevention of scum layer formation when properly mixed.
(5) Even temperature distribution throughout the substrate.
(6) Maintenance of minimum levels of inhibiting substances at any one point in the digester.
(7) Avoids plugging, gas entrapment, and channeling.
(8) Easily modeled.

Disadvantages include:

(1) Large reactor requirements due to the inability of this digester to operate with a mean SRT and MRT greater than the HRT.
(2) Large power requirements to operate adequate mixing mechanisms.
(3) Difficult to achieve complete mixing at commercial scale.
(4) Incompletely digested substrate leaving the system.
(5) No separation of acid-forming bacteria and methane-producing bacteria, thereby limiting the optimum environmental requirements of the different bacteria.
(6) Loss of microorganisms with the digester effluent.

CSTRs are used widely and are considered conventional for sewage sludge digestion (that is, high-strength organic wastes). As previously indicated, long SRTs and MRTs also require long HRTs and result in large reactor sizes and higher capital costs. The conventional stirred tank digestion process has not been proven to be economically feasible for treatment of wastes containing less than 1000 mg/liter BOD.[12] This is because of the large reactor volume required for low-strength organic feeds.

The high SRTs required to break down significant portions of the cellulose, hemicellulose, and lignin content of biomass feeds, as reported in the discussion on hydrolytic bacteria, will also result in large HRT requirements. In addition, mixing of feeds containing biomass may be difficult due to paste-like consistencies, a range of particle sizes, or other heterogeneous feed properties resulting from a mixture of plant components. Prevention of scum layer formation also may not be possible at normal mixing speeds.

9.3.1.3 Plug-Flow Reactors (PFRs)

These reactors, illustrated in Fig. 9.6, typically receive feed at one end and remove effluent from the other. Such reactors are not mechanically mixed and, ideally, the biomass feed passes through the reactor in a 'plug-like' manner. This means that the liquid, solid, and microorganism portions of the biomass enter and exit the reactor at approximately the same time. Although no mechanical mixing is provided, some vertical mixing of the digester contents does occur during gas production. However, longitudinal mixing is minimal, thereby promoting phase separation. The passive settling of solids, therefore, results in longer SRTs than HRTs. In this type of reactor scheme, feedstock hydrolysis and acid production occur near the front portion of the digester, while methane production occurs nearer the effluent end. Unless microorganisms are continuously seeded into the reactor, as with manure feeds, the bacteria needed to perform the functions of anaerobic digestion are removed from the digester, causing decreased stability and eventual failure.

Advantages of this digestion scheme are:

(1) Simplicity and reduced energy input, as mechanical mixing is not required.
(2) Very synergistic with manure characteristics and, therefore, often most economical for farm applications.
(3) High stability.
(4) Best suited for well-suspended, high-solids wastes.

Disadvantages are:

(1) Solids may settle out, reducing the effective reactor volume and HRTs and SRTs.
(2) Need for solids/microorganism recovery.[13]
(3) Lower efficiencies for cold feedstocks requiring substantial preheating to mesophilic or thermophilic temperatures because of high surface-to-volume ratios.
(4) Difficult to maintain uniform temperatures and substrate conditions.

(5) Formation of thick crusts.[13]

Plug-flow systems may be suitable for some crop residue or grassy feedstocks provided an inoculum can either be maintained in or routinely added to the system.[6] Biomass feedstocks that would have a tendency to settle out of suspension would not be suitable.

Laboratory studies of plug-flow digesters operating at near ambient temperatures show gas yields about one-third those of conventional digesters, but the lower temperatures make it possible to compare net gas yields with the lower energy requirements to heat the digester and mix its contents.[14] Some biomass application may exist if the feedstock cost is low or negligible. Plug-flow reactors can handle very concentrated wastes and semi-solid materials — that is, over 6% volatile solids — if material properties are such that they do not accumulate in the reactor. This could be advantageous to biomass digestion.

Vertical baffles have been installed in some digesters to create dead space for harboring active organisms, thereby reducing washout. Such a baffle-flow digester was demonstrated on kelp and on soluble feedstocks. Most biomass feeds will require solids recycle or baffle-flow construction to maximize retention of feed solids and microorganisms in the digester.

9.3.2 Reactor Category B

Examples of Category B include continuously stirred tank reactors with solids recycle (CSTR/SR), upflow solids reactors (USR), and upflow sludge blanket reactors (USBR). These reactors promote long SRTs and MRTs either actively, through dewatering techniques, or passively, by settling. The methane production rate in these reactors can be increased by increasing the feed loading rate while maintaining the methane yield and promoting long SRTs and short HRTs. Digesters that promote short HRTs while maintaining long SRTs can process larger amounts of feed without increasing the reactor size. These reactors are suited for feeds with low to high solids content that require long SRTs for efficient conversion.

9.3.2.1 Continuously Stirred Tank Reactor With Solids Recycle (CSTR/SR)

The CSTR/SR illustrated in Fig. 9.6 is a CSTR that promotes the retention of microorganisms and unreacted solids by settling and recycling of the effluent solids into the stirred tank. This then increases the system's SRT over the HRT resulting in reduced reactor volumes, higher loading rates, and more favorable economics. Therefore, organic feeds that are relatively soluble can be treated at lower HRTs while maintaining the SRT required for the slow-growing methanogenic bacteria.

The system maintains all of the advantages of the CSTR with the additional benefit of increased SRTs and smaller reactor volumes. However, disadvantages may result from the additional cost of the settling basin, recycle pumps, and heat losses in the recycle loop. Also, a solids/gas separator may be necessary to effect proper settling of the solids and, therefore, must be included in the design costs.

The settling characteristics of the biomass digester effluent will affect the suitability of this reactor design for specific applications. In addition, if the biomass feed contains a significant quantity of inert material, such as ash or grit, then a recycle system could reduce the effectiveness of the digester by a build-up of the inert fraction in the digester.[15] Doyle *et al.*,[16] for instance, are currently investigating various methods of improving the economics of producing methane from corn stover using a CSTR/SR. In addition to using cell recycle for microorganism recovery and culture enhancement, they successfully increased the feed solids concentration to 20% while maintaining a 10% reactor solids concentration.

9.3.2.2 Upflow Anaerobic Sludge Blanket Reactors (UASBR)

The UASBR illustrated in Fig. 9.6 features a bed of granular and flocculated sludge suspended by an upflow of wastewater. The reactor consists of three distinct zones: a sludge bed, a sludge blanket, and a settling/gas separation zone. Wastewater feeds into the tank from below and flows upward through a bed of dense, granular sludge and a blanket of flocculated sludge particles. As the liquid moves through the floc, increased MRTs and SRTs are achieved relative to the HRT. Gas and suspended solids are separated from the treated liquid by means of an internal gas–solids separator device. Baffles are used to release trapped gas, thus promoting the settling of sludge particles back toward the digesting zone, promoting long SRTs and high concentrations of sludge in the digester.

Although gas bubbles from the produced gas provide mixing, some of the influent may bypass the sludge-bed zone via cracks and channels. The sludge-bed zone has been reported to be responsible for 80–90% of the waste degradation in the reactor and to occupy approximately 30% of the reactor volume.[12] After passing through the sludge-bed zone, the influent enters the sludge-blanket section, which occupies the remaining 70% of the reactor volume. With the solids concentration in this zone lower than that of the sludge bed, rising gas bubbles mix the flocculated sludge blanket. The need for an internal combination solids-settler/gas-collector device adds an additional expense in the construction of this type of digester.

This process is often considered to be an advanced form of waste treatment for high-strength organic influents.[17]

Advantages of this system include:

(1) Simple construction with the exception of the gas/solids separator.
(2) Higher loading rates are obtainable.
(3) Low concentrations of suspended solids possible in influent.
(4) No mechanical mixing or costly support media required.
(5) Low effluent suspended solids.
(6) Improved environmental conditions for the microorganisms by separating acid and methane bacteria.

Disadvantages are:

(1) Requires effective gas and liquid separator.
(2) Needs efficient means of distributing the feed over the bottom of the tank.
(3) May lose solids and microorganisms in the effluent at high solids loading rates.
(4) May lose a portion of the sludge bed during a hydraulic surge or toxic upset.
(5) Requires effluent recycle for bed expansion.

9.3.2.3 Upflow Solids Reactor (USR)

The USR, illustrated in Fig. 9.6, is a simple, low-cost design concept that passively promotes higher SRTs and MRTs relative to the HRTs. Unreacted biomass solids and microorganisms are maintained in the reactor by passive settling. Using particulate biomass feeds, effluent recycle and gas/solids separators are not required.[18] The dense accumulation of solids in the bottom of the digester promoted by a relatively high length-to-diameter ratio (typically between 1:2 and 2:1) results in high SRTs.

The applicability of a mesophilic USR digester to sea kelp, with a total solids concentration of approximately 12%, was studied.[19] At loading rates ranging from 1·6 to 9·6 kg VS/m^3-day, the methane yield ranged from about 0·38 to 0·34 m^3/kg added, and the methane production rate ranged from 0·6 to 0·32 vol/vol culture-day, respectively. The observed methane yield exceeded 70% of theoretical and the overall performance of the USR was significantly better than that of a CSTR.

That study showed that high methane yields, methane production rates, and good process stability could be achieved with a 12% total solids concentration feed by using a simple digester design that promoted long SRTs. Previous applications of a similar reactor design concept (the upflow sludge blanket reactor) had been restricted to soluble feeds. These findings extended the application of this design concept to feedstocks with high suspended solids content.

9.3.3 Reactor Category C

In this category the emphasis is placed on retaining the microorganisms in the reactor while permitting the solids and the liquid portions of the biomass feed to pass through the system. These reactors contain a solid support medium, such as sand, charcoal, plastic, pall rings, or rocks, upon which bacteria attach, thus preventing their washout at relatively short HRTs. Consequently, these reactors are best suited for highly biodegradable feeds containing relatively low concentrations of particulate-associated solids that require short SRTs for effective conversion. Examples of attached film reactors include the anaerobic filter (AFR), fluidized-bed (FBR), and expanded-bed (EBR) reactors.

9.3.3.1 Anaerobic Filter Reactors

The AFR, shown in Fig. 9.6, consists of a filter bed filled with inert support media such as gravel, rocks, charcoal, or plastic. The stationary support media may be random or oriented and the liquid flows either up or down. Bacteria attach to the filter media as a film and absorb nutrients from the flow of influent. As this film accumulates, a portion sloughs off and passes out of the digester as sludge. Porous surfaces promote bacterial attachment. Specially designed media, such as plastics and bricks, have the advantage that they can be shaped to achieve higher surface-to-volume ratios.

AFRs tend to separate the different microbial populations. With upward flow, the acid-forming bacteria act on the influent at the bottom of the digester, while the methane formers are found attached to the bed further along at locations more distant from the influent. The bacteria remain stationary as the byproducts move through the bed. The packed filter medium not only retains bacteria but also provides a mechanism for the separation of solids and produced gas. This type of system results in rapid-rate biomethanogenic conversion of soluble or fully biodegradable wastes. Feeds with high suspended solids levels are generally not suitable as they rapidly clog the system. The use of oriented packing rather than random packing and downflow operation is believed to reduce plugging problems.[20] Another method used to prevent plugging and dead spaces in the digester is to feed the influent from the top. This downflow scheme is being used, for instance, in the Bacardi digester on rum distillery wastes.[21] However, such flow patterns may also lead to increased microorganism washout.

Advantages of a media-packed reactor over the CSTR with recycle include:[21,22]

(1) Lower operating costs; that is, no mixing required.
(2) Smaller tank volume because of higher loading rates.
(3) More stable system because of the high concentrations of organisms.
(4) More able to absorb load variations.
(5) Longer retention of microbes because of the fixed habitat.
(6) More rapid restart after prolonged shutdowns.

Disadvantages are:[21,23]

(1) Higher costs of packing material (reported at up to 60% of total cost).
(2) Increased pressure drop during operation due to biomass accumulation.
(3) Greater possibility for plugging digester with suspended solids.
(4) Longer start-up periods usually required.

As this type of reactor is not suited for influents containing suspended, particulate solids, it is unlikely that it will find direct application to high-particulate-solids biomass feeds. However, laboratory investigations are being conducted at the Institute of Gas Technology (IGT) and other laboratories using packed-bed digesters for the methane phase component of two-phase digestion.[23] Only the solubilized portion of the influent would then feed to the digester, freeing it from potential plugging problems.

Extensive research has been done at the laboratory scale using vegetable, dairy, piggery, and distillery wastes as feeds.[24–26] The National Research Council of Canada has been involved in research using a downflow stationary fixed film reactor. COD removal ranged from 27%

to 95%. Pilot plant studies have been conducted on a variety of wastewaters including dairy wastes, raw municipal wastes, piggery wastes, and confectionery wastes.[23,25,27,28]

9.3.3.2 Fluidized- and Expanded-Bed Reactors

The fluidized- and expanded-bed reactors (FBR and EBR) promote longer MRTs than HRTs. Wastewater is passed up through a bed of small (0·3–50 mm) inert (for example, sand) or reactive (for example, carbon) particles, referred to as bacterial support media, at sufficient velocities partially to expand or to fluidize the bed, as illustrated in Fig. 9.6. In a properly operated fluidized-bed digester, sufficient hydraulic mixing is provided to suspend completely the support medium. An expanded-bed digester, on the other hand, only provides sufficient hydraulic flow to cause an expansion of the support medium from its settled state (thereby decreasing the density) without achieving a completely mixed state.

In either case, the support medium provides a large surface area that supports the growth of attached microorganisms. These microorganisms are retained in the reactor while the solids and liquid portions of the feed are permitted to pass through the system. The bacteria are retained for long periods, even at relatively short HRTs. Increasing the surface area available for growth allows for higher loadings and provides greater resistance to toxics.[20]

FBR and EBR digesters are best suited for highly biodegradable feeds containing low concentrations of particulate-associated solids that require short SRTs for effective conversion. They have two major advantages over packed-bed systems: (1) larger surface areas are available for attachment of organisms, and (2) some particulate feeds can pass through the support media. A major drawback is greater energy use, since recycle ratios of 0·5–10 or higher may be needed to achieve fluidization.[20]

The advantages of this system, therefore, are:

(1) Higher microorganism concentrations due to the large specific surface area available for attachment.
(2) High loading rates are attainable.[29]
(3) More stable system because of the high concentration of minor microorganisms.
(4) Load variations can be absorbed.
(5) More rapid restart after prolonged shutdowns.
(6) Low concentrations of particulate-containing feeds can be used.
(7) Promote good mixing.

Disadvantages are:

(1) May require high energy and maintenance cost for fluidization.
(2) Support media may wash out, damaging pumps and other equipment.[30]
(3) Potentially high cost to recover media lost in effluent.
(4) Cannot accept feeds containing high concentrations of particulate-associated solids.
(5) Requires long start-up periods.
(6) No separation of acid and methane phase bacteria.
(7) Degasifier may be required to adequately separate the media and solids from the effluent.

It is unlikely that this type of digester will be very effective for biomass applications requiring long biomass SRTs. The hydrolysis of the ligno-cellulose fractions of biomass feeds, for example, requires long SRTs that are not effectively provided in this type of digester. In addition, the physical characteristics of biomass feeds (paste-like or highly particulate) may result in excessive washout of the support media requiring large settling basins. If not adequately fluidized or expanded, plugging, channeling, and other effects of maldistribution of the inventory (that is, temperature control) can occur.

The high concentration of attached microorganisms can be used to treat effectively highly biodegradable soluble feeds. Also, investigations are being made in the laboratory to utilize FBRs, as methane phase digesters in two-phase digestion schemes.[23] It may also be possible to extract organic liquors from biomass and then treat this liquor in a fluidized-bed digester. The cellulose fraction would then be treated separately in another digester optimized for long SRTs.

9.3.4 Two-Phase Digestion

As discussed earlier, anaerobic digestion is performed by distinct groups of bacteria that differ significantly with respect to physiology, nutritional requirements, growth characteristics, metabolic characteristics, environmental optima, and sensitivity to environmental stress.[31] Although some digester concepts such as the PFR and USR digesters promote separation of bacterial groups, it is very difficult to optimize conditions for different groups of bacteria in one reactor.

By taking advantage of the differences in the relative growth rates of methanogenic and hydrolytic/acidogenic bacteria, dominant populations of either of these two bacterial groups can be promoted by controlling digester operating conditions. As previously discussed, methanogenic bacteria grow more slowly than those in the nonmethanogenic group. Consequently, operating digesters at short HRTs and MRTs promotes the selective washout of methanogens. Such a digester will promote hydrolysis of the biomass feed into simpler components that can then be metabolized into fermentation products, such as volatile acids. Because methanogenic bacteria are absent from this system, little methane production occurs.

The fermentation products produced in the hydrolytic/acidogenic digester can be added to a second digester in which a large methanogenic bacterial population is promoted by operation at long MRTs. In this methane-phase digester, the fermentation products are used as substrates for methane production. With proper operation, significantly higher methane than carbon dioxide concentrations will be produced in this methane phase. Most of the carbon dioxide is produced in and discarded from the nonmethanogenic phase.

The two-phase concept can be employed with several reactor configurations, provided the MRT in the second phase exceeds that in the first phase digester. Figure 9.6 illustrates a two-phase system in which a CSTR is employed as the nonmethanogenic phase and an AFR as a methanogenic phase. The choice of the specific reactor combinations is determined by feedstock characteristics as well as by interactive dynamics that affect reactor compatibility.

Besides the inherent advantages of individual digesters, two-phase digestion has the following additional advantages:[6]

(1) Ability to maintain appropriate densities of acid and methane formers.
(2) Maximization of conversion rates through independent control of temperature, pH, oxidation–reduction potential, biomass recycle, retention time, and other parameters for each phase.
(3) Enrichment of the product gas with methane.
(4) Greater stability with respect to feedstock loading, pH, and toxic shock.
(5) Higher solids reduction at lower retention time and, therefore, reduced total volume per unit of influent processed.
(6) Reduced gas clean-up costs where required, due to the increased methane content of the product gas.

The main disadvantage is the cost of the more complex system. Capital expenditure for the additional equipment, controls, and labor requirements may be higher than with conventional systems.

In application to biomass feedstocks, the separation of the hydrolytic/

acidogenic and methanogenic phases is likely to derive the same benefits observed with other feedstocks. Different reactor configurations can be chosen to optimize reaction rates and to take advantage of changes in physical properties between phases. For high-solids biomass feeds, a leachate-bed recycle system (consisting of a hydrolytic/acidogenic-phase digester for the biomass solids and an attached film, methane-phase digester for the leachate) might be a very desirable combination. The selection of the individual digester concepts for each phase and their inherent advantages and disadvantages are likely to have the highest impact on the effectiveness of the overall process. Considerably more data and experience are needed on individual digester performances for biomass feedstock in order effectively to evaluate two-phase system benefits.

9.3.5 Other Digestion Schemes

The aforementioned digesters were chosen for discussion because they represent major categories covering separate and distinct principles of operation. Within each category, hybrid versions of these concepts can be found and may be considered for biomass application. As discussed earlier, the baffle flow reactor, a form of the plug-flow digester concept, has been tested on kelp feedstock and on intermediate- and low-strength, soluble industrial wastes.[32,33] Anaerobic rotating biological discs have also been applied to waste treatment but may be too costly and complex in construction for many applications. Other variations result from the use of auxiliary equipment such as the use of degasifiers, thickeners, and membrane technology[34] to improve the effectiveness of an application. Other concepts and techniques that are under study[10,35,36] show similar potential for biomass conversion; however, the limits of this review prohibit a more extensive presentation.

9.4 REACTOR SELECTION FOR BIOMASS FEED CATEGORIES

9.4.1 Soluble Feeds

Soluble feeds are typically highly biodegradable, containing low concentrations of particulate-associated solids that require short HRTs and SRTs for conversion to methane. The CSTR has been widely used for the digestion of low- to high-strength soluble feeds, such as sewage sludge. This reactor type is operated as a completely mixed system in which the soluble feed HRT is equal to the particulate feed SRT and the MRT.

Soluble feeds require minimal hydrolysis and can therefore be digested at short HRTs. However, at short HRTs higher numbers of microorganisms will wash out of the reactor causing poor digester performance.

Shorter HRTs can be achieved without affecting digester performance by using reactors such as anaerobic filter (AFR) or upflow anaerobic sludge blanket reactors (UASBR), which effectively retain microorganisms. Using synthetic soluble feeds increased methane production rates in AFRs and USBRs were attributed to longer SRTs together with higher loading rates.[9] Good performance has been reported with UASBRs treating soluble dairy and food processing wastes,[9] and with an AFR for treating wastewater.[19] Colleran *et al.* achieved COD reductions of 80–88% with soluble piggery, milk, and silage effluent wastes in an AFR at HRTs ranging from 0·5 to 3 days.[37] The observation of improved digestion with two-phase digestion in comparison to a CSTR[38] led to the development of a commercial process.

9.4.2 Low to Medium Solids

Low- to medium-solids feeds typically consist of a highly degradable soluble fraction and a particulate-associated fraction that is more resistant to digestion. In addition to retaining microorganisms, it is necessary to retain the particulate fraction of feed, requiring reactors that achieve longer SRTs and MRTs in comparison to HRTs. By removing the soluble fraction of the feed from the digester at a faster rate than the particulate fraction, longer SRTs can be obtained without increasing the reactor size. Examples of reactors designed for these feeds are CSTR/SRs, USRs, and USBRs.

A USR demonstrated increased performance for biomass feeds with less than 15% total solids concentrations. Fannin *et al.*[18] found that a USR exhibited higher methane yields and process stability than a CSTR during the digestion of sea kelp. The digestion of water hyacinth/sludge blend in a USR also demonstrated higher methane yields than a CSTR operated under similar conditions.[39] Another approach examined for the digestion of biomass at medium- to high-solids concentrations was a two-stage process consisting of liquefaction and gasification. Marine and freshwater algae were converted to methane by a two-step anaerobic digestion process. The first step consisted of a percolation step for liquefaction of the algae followed by an upflow methane digester for conversion of the soluble products.[40] This approach was used for the digestion of wheat straw and grass by Colleran *et al.*[37] The first-stage reactor was operated as a batch-fed percolation system to promote hydrolysis of the biomass. The reactor was initially fed dry feed (>25% solids concentration). Optimum liquefaction, however, occurred at 10–15% solids concentration. Biomass (wheat straw) and solid wastes were digested using a two-stage process in which water was percolated through a batch-fed reactor, followed by conversion of the dissolved organics to methane in a USBR.[41] Leachate from a bark landfill was treated by using the landfill as a percolator. The leachate was introduced below the upper surface of the landfill with continuous recycling and pH control to develop a methane-producing flora.[42] IGT has demonstrated that high methane yields can be achieved from herbaceous and woody biomass; however, long SRTs are needed for degradation of high-lignin-content feeds. They have suggested the use of a two-stage interactive system to achieve these high yields at reduced SRTs.[43]

9.4.3 High Solids

The majority of land-based biomass feeds, as harvested, have a total solids concentration greater than 15%.[43] Anaerobic fermentation of these feeds with minimum water addition permits higher organic loadings and smaller digester volumes when compared with soluble feed fermentation systems. A greater net energy production efficiency can be achieved because less process energy is required to heat the digester feed.

Batch-fed combined-phase and two-phase percolation systems have been studied for the conversion of high solids biomass to methane.[9,11,44] These percolation systems were initiated as high solids reactors using water-saturated feeds ranging from 10% to 15% TS. Data from batch-fed combined phase fermentation systems studied for the conversion of dry agricultural residues (>20% TS) to methane indicated that increasing the total solids content slowed down the anaerobic fermentation.[11] Methane production became limited at a 32% solids concentration while hydrolysis continued until the solids concentration reached approximately 60%. Batch conversion of 90% of the biodegradable matter was achieved at an initial TS concentration of 28% in 200 days. The biodegradable fraction, however, ranged from 30% to 60% of feed VS, depending on the incubation conditions. Larger-scale reactor studies showed approximately 48% VS reduction following a 280-day incubation period; however, the TS content of the wheat straw substrate was only 18%. Additional water was added during the study to provide sufficient moisture for reactor heating. Corn stover was successfully digested in a semicontinuously fed STR at a TS concentration of 20%. Conversion efficiencies of 66% were achieved at a 60-day retention time.[16]

9.5 SUMMARY

The major digester types under development or in use today and their potential use for biomass gasification are reviewed. A single reactor configuration and set of operating conditions may not be suitable for the anaerobic digestion of all forms of biomass. Factors such as feed solids content and reactor configuration can affect the SRT, which, in turn, can be a critical parameter affecting digester nutrient requirements, biodegradability, and methane yields. The solids content of different feeds also determines the HRT and the feed heating requirements of anaerobic digestion.

Longer SRTs and MRTs promote greater biodegradation of feeds, enhance digester stability, and reduce the nutrient requirements for digestion. Longer SRTs and MRTs can be promoted in several reactor designs without increasing the HRT. However, because there are significant differences in the chemical and physical properties of many biomass feeds, optimal anaerobic digester performance requires matching feed characteristics with reactor configurations.

Highly biodegradable feeds with low solids concentrations can be gasified efficiently in attached film reactors. On the other hand, feeds with high particulate-associated solids concentrations are not very suitable for use in attached film reactors because of operational and mechanical problems such as clogging of the microorganism support media. Less biodegradable feeds with high particulate-associated solids concentrations and relatively low nutrient concentrations can be more effectively digested in reactors, such as the USR, that promote long SRTs and MRTs.

The physical and chemical characteristics of biomass feeds have an important impact on the methane yields and production rates from anaerobic digestion. Since no single digester design or set of operating conditions can be applied to all biomass types, a limited number of prototype species should be selected on the basis of their particulate solids content and chemical composition to serve as a basis for intense investigation. Research with respect to digester design and operating conditions should be addressed specifically to the physical and chemical composition of biomass. The results of these studies should be translatable to other species with similar composition. These studies should also consider the effects of growth and harvest conditions on interspecies variability of chemical composition that can impact digester performance.

Work should continue to evaluate the effects of SRT on digester stability and overall performance. Concepts such as nutrient and cell synthesis requirements should be reevaluated under conditions of long SRTs. The conditions that promote long MRTs in reactors should also be evaluated. These studies might address such questions as factors promoting attachment of microorganisms to both inert and biomass substrates.

Research on anaerobic digestion should be directed toward improving loading rates and methane yields. These studies should clearly differentiate the relative importance of solids, microorganism, and hydraulic retention time. Hydraulic retention should be deemphasized as a parameter for evaluating digester operation. Research on biomass for energy should be based on major principles that have been developed in the bovine rumen and sewage sludge digestion. However, these principles should be reevaluated before application to biomass digestion.

REFERENCES

1. Fannin, K. F., Srivastava, V. J., Mensinger, J. D. and Chynoweth, D. P., 'Marine Biomass Program: Anaerobic Digestion Systems Development and Stability Study,' IGT Project *30558* Final Report, Chicago, June 1983.
2. Pfeffer, J. T., 'Increased Loadings on Digesters With Recycle of Digested Solids.' *J. Water Poll. Cont. Fed.* **40**:1920–1933, 1968.
3. Fannin, K. F., Srivastava, V. J., Chynoweth, D. P. and Bird, K. T., 'Effects of the Interaction Between Biomass Composition and Reactor Design on Anaerobic Digestion Process Performance,' in Symposium Papers, *Energy From Biomass and Wastes VII.* Chicago: Institute of Gas Technology, 1983.
4. Boening, P. H. and Larsen, V. F., 'Anaerobic Fluidized Bed Whey Treatment,' *Biotechnol. Bioeng.* **24**:2539, 1982.
5. Speece, R. E. and McCarty, P., 'Nutrient Requirements and Biological Solids Accumulation in Anaerobic Digestion,' in *Adv. Water Poll. Res.* **9**, Vol. 2. New York: Pergamon Press, 1964.
6. Abeles, T., Ellsworth, D. and Genereux, J., *Task VI Biological Production of Gas for Office of Technology Assessment*, Washington, DC, April 1979.
7. Lettinga, G. and Vinken, J., 'Feasibility of the Upflow Anaerobic Sludge Blanket (USAB) Process for the Treatment of Low-Strength Wastes,' *Proc. 35th Ind. Waste Conf.*, 625, Purdue University, 1980.
8. Wise, D. L., *et al.*, '*In Situ* Methane Fermentation of Combined Agricultural Residues,' pp. 275–294 in *Resources and Conservation* **6**. Amsterdam: Elsevier Scientific Publishing Company, 1981.
9. Heertjes, P. M. and van der Meer, R. R., 'Comparison of Different Methods for Anaerobic Treatment of Dilute Wastewaters,' *Proc. 34th Ind. Waste Conf.*, 790, Purdue University, 1979.
10. Jewell, W. J., *et al.*, 'Agricultural and High Strength Wastes,' pp. 255–268 in

Anaerobic Digestion 1981, Proceedings of the Second International Symposium on Anaerobic Digestion. Travemunde, Federal Republic of Germany, 1981.

11. Jewell, W. J., Cummings, R. J., Dell'Orto, S., Fanfoni, K. J., Fast, S. J., Gottung, E. J., Jackson, D. A. and Kabrick, R. M., 'Dry Fermentation of Agricultural Residues.' Final Project Report for SERI, *SERI/STR-231-1892*, 1982.
12. Obayashi, A. W. and Gorgan, J. M., 'Management of Industrial Pollutants by Anaerobic Processes,' prepared for the Industrial Waste Elimination Center.
13. Chandler, J. A., Hermes, S. J. and Smith, K. D., 'A Low Cost 75 kW Covered Lagoon Biogas System,' in Symposium Papers, *Energy From Biomass and Wastes VII*. Chicago: Institute of Gas Technology, 1983.
14. Stafford, D. A. and Hughes, D. E., 'Fermentation to Biogas Using Agricultural Residues and Energy Crops,' pp. 406–410 in *Proceedings of the Industrial Conference on Biomass*, Brighton, England, 'Energy From Biomass,' Brighton, England, 4–7 November, 1980.
15. Pfeffer, J. T., 'Engineering, Operation and Economics of Methane Gas Production,' *Experientia* **38** (1982), Birkhauser Verlag, CH-4010, Basel/Switzerland.
16. Doyle, O. P., O'Malley, J. D., Clausen, E. C. and Gaddy, J. L., 'Kinetic Residues,' in Symposium Papers, *Energy From Biomass and Wastes VII*, 545–561. Chicago: Institute of Gas Technology, 1983.
17. Obayashi, A. W., Stensel, H. D. and Komirek, E., 'Anaerobic Treatment of High-Strength Wastes,' *Chem. Eng. Progr.* **77**:68–73, April 1981.
18. Fannin, K. F., Srivastava, V. J. and Chynoweth, D. P., 'Unconventional Anaerobic Digester Designs for Improving Methane Yields From Sea Kelp,' in Symposium Papers: *Energy From Biomass and Wastes VI*, 373–400. Chicago: Institute of Gas Technology, 1982.
19. Genung, R. K., *et al.*, 'Energy Conservation and Scale-Up Studies for a Wastewater Treatment System Based on a Fixed-Film Anaerobic Bioreactor,' *Biotechnol. Bioeng. Symp.* **10**:295, 1980.
20. Olthof, M. and Oleszkiewicz, J., 'Anaerobic Treatment of Industrial Wastes,' *Chem. Eng.* 121–126, 1982.
21. Szendrey, L. M., Schafer, P. E. and Dorion, G. H., 'Pollution and Energy Management Through the Anaerobic Approach,' *Industrial Wastes* 31–34, 1982.
22. Szendrey, L. M., 'The Bacardi Corporation Digestion Process for Stabilizing Rum Distillery Wastes and Producing Methane,' in Symposium Papers, *Energy From Biomass and Wastes VII*. 24–28. Chicago: Institute of Gas Technology, 1983.
23. Genung, R. K., Harris, M. T., Rivera, A. L. and Donaldson, T. L., 'Operation of an Upflow Fixed-Bed Anaerobic Digester for Waste Stabilization and Fuel Gas Production at Near-Commercial Scale,' in Symposium Papers, *Energy From Biomass and Wastes VII*, 24–28. Chicago: Institute of Gas Technology, 1983.
24. Van den Berg, L. and Kennedy, K. J., 'Effect of Substrate Composition on Methane Production Rates of Downflow Stationary Fixed Film Reactors,' in

Symposium Papers, *Energy From Biomass and Wastes VI*, 401–420. Chicago: Institute of Gas Technology, 1982.

25. Colleran, E., Barry, M., Wilkie, A., 'The Application of the Anaerobic Filter Design to Biogas Production from Solid and Liquid Agricultural Wastes,' in Symposium Papers, *Energy From Biomass and Wastes VI*, 443–478. Chicago: Institute of Gas Technology, 1982.
26. Stevens, T. G. and Van den Berg, L., 'Anaerobic Treatment of Food Processing Wastes Using a Fixed-Film Reactor,' *Proc. 36th Ind. Waste Conf.*, 224–232, Purdue University, 1981.
27. Wheatley, A. D. and Cassell, L., 'Energy Recovery and Effluent Treatment of Strong Industrial Wastes by Anaerobic Biofiltration,' in Symposium Papers, *Energy From Biomass and Wastes VII*. Chicago: Institute of Gas Technology, 1983.
28. Colleran, E., *et al.*, 'Bioconversion of Agricultural Wastes into Fuel Gas and Animal Feed,' pp. 416–422 in *Energy From Biomass 1st E.C. Conference*, W. Plaz, P. Chartier, and D. O. Hall, Eds, London: Elsevier Applied Science Publishers, 1981.
29. Sutton, P. M., Li, A., Evans, R. R. and Korchin, S. R., 'Dorr–Oliver's Fixed Film, Suspended-Growth Anaerobic Systems for Industrial Wastewater Treatment and Energy Recovery,' pp. 667–675 in *Proceedings of the 37th Industrial Waste Conference*, Purdue University, 1982.
30. Miller, W. D., 'Anaerobic Pretreatment and Soft Drinks,' paper presented at the Symposium *Anaerobic Biotechnology: Reducing the Cost of Industrial Wastewater Treatment*,' Arlington, Va, 1983.
31. Stafford, D. A., Hawkes, D. L. and Horton, R., *Methane Production From Waste Organic Matter*. Boca Raton, Fla: CRC Press, 1980.
32. Chynoweth, D. P., Srivastava, V. J. and Conrad, J. R., 'Research Study to Determine the Feasibility of Producing Methane Gas From Sea Kelp,' Annual Report for General Electric Company, prepared by the Institute of Gas Technology, April 1980.
33. Bachmann, A., Beard, V. L. and McCarty, P. L., 'Comparison of Fixed Film Reactors With Modified Sludge Blanket Reactor,' Paper presented at the 4th International Conference on Fixed Film Biological Processes, Kings Island, Ohio, 20–23 April, 1982.
34. Choate, W. T., Houldsworth, D. and Butler, G. A., 'Membrane-Enhanced Anaerobic Digesters,' pp. 651–660 in *Proceedings of the 37th Industrial Waste Conference*, Purdue University, 1982.
35. Hashimoto, A. G., 'Conversion of Straw–Manure Mixtures to Methane at Mesophilic and Thermophilic Temperatures,' pp. 185–200 in *Biotechnology and Bioengineering*, Vol. XXV. New York: John Wiley & Sons, 1983.
36. Hashimoto, A. G. and Robinson, S. A., 'Pilot-Scale Conversion of Manure–Straw Mixtures to Methane,' pp. 19–28 in *Resources and Conservation* **6**. Amsterdam: Elsevier Scientific Publishing Company, 1982.
37. Colleran, E., *et al.*, 'The Application of the Anaerobic Filter Design to Biogas Production From Solid and Liquid Agricultural Wastes,' p. 43 in Symposium papers, *Energy From Biomass and Wastes VI*, D. L. Klass, Ed., Chicago: Institute of Gas Technology, 1982.
38. Ghosh, S. and Henry, M. P., 'Stabilization and Gasification of Soft-Drink

Manufacturing Waste by Conventional and Two-Phase Anaerobic Digestion,' *Proc. 36th Ind. Waste Conf.*, Purdue University, 1981.

39. Chynoweth, D. P., Dolenc, D. A., Ghosh, S., Henry, M. P., Jerger, D. E. and Srivastava, V. J., 'Kinetics and Advanced Digester Design for Anaerobic Digestion of Water Hyacinth and Primary Sludge,' *Biotech. and Bioeng. Symp.* No. **12**:381–398, 1982.
40. Asinari, Di San Marzano, C.-M., *et al.*, 'Methane Production by Anaerobic Digestion of Algae,' *Energy From Biomass* **1**:13, 1981.
41. Rijkens, B. A., 'A Novel Two-Step Process for the Anaerobic Digestion of Solids Wastes,' in Symposium Papers, *Energy From Biomass and Wastes V*, D. L. Klass and J. W. Weatherly, III, Eds., Chicago: Institute of Gas Technology, 1981.
42. Rydin, S. and Haglund, P., 'Control of Bark Landfill Leachates With an Anaerobic Treatment System,' *TAPPI* **64**:59, 1981.
43. Chynoweth, D. P., Jerger, D. E., Conrad, J. R., Razik, A., Srivastava, V. J., Ghosh, S., Henry, M. P. and Babu, S., 'Gasification of Land-Based Biomass,' Final Project Report for IGT Project No. *30546*, Chicago, 1983.
44. Chynoweth, D. P., Dolenc, D. A., Swegler, B. and Reddy, K. P., 'Wastewater Reclamation and Methane Production Using Water Hyacinth and Anaerobic Digestion,' Paper presented at the *10th Energy Technology Conference*, Washington, DC, February 1983.

10
Start-Up, Operation, Stability, and Control

KIRBY F. FANNIN
Institute of Gas Technology, Chicago, Illinois, USA

10.1 INTRODUCTION

Digester operation requires successful start-up and assurance that performance is within acceptable limits with respect to the efficiency of biodegradation, methane production rate, stability, and reliability of the process. Such assurance requires determination of digester performance using several physical and chemical parameters as a basis for evaluation and control. The temperature of digestion, mixing conditions and techniques, methods and frequency of feeding, and pressure within the digester are among the operational variables that help determine digester performance and stability. After the operational methods have been defined, methods for detecting digester imbalance and restoring and maintaining stability are needed. The purpose of this chapter is to discuss start-up and operating techniques that affect digestion, causes of digester stability, and methods for controlling digester imbalance and promoting stability.

10.2 START-UP

Specific methods for starting anaerobic digesters may vary somewhat, depending upon the reactor design and digesting substrate. The general start-up procedure initially involves addition of a seed culture, or inoculum, containing the microorganisms required for digestion to the reactor. The biomass feedstock is then added to the culture at a low loading rate in order to ensure stable digester performance. As the microorganisms in the culture grow, a balance between the groups of organisms performing several metabolic functions is maintained. The biomass loading rate is gradually increased until the planned feed loading is achieved. Successful digester start-up requires utilization of a viable anaerobic culture containing a broad spectrum of microorganisms and controlled introduction of biomass feed. A properly functioning digester maintains a healthy microbial population and does not normally need the continuous addition of inoculum. The organisms, once seeded into the digester, reproduce and utilize available substrate. The rate of start-up is affected by the characteristics of the inoculum, the biomass feed, the configuration and size, and operating conditions of the anaerobic digestion reactor.

10.2.1 Inoculum and Feed

Anaerobic digestion is performed by a consortium of different species of bacteria that perform functions of biomass hydrolysis, degradation of these hydrolytic components to fermentation products such as volatile acids, and utilization of these fermentation products for the production of methane and carbon dioxide. The failure of an anaerobic digestion process to degrade certain biomass feed components may be attributed to the absence of organisms capable of their degradation under the conditions of digestion. However, certain biomass components, such as lignin, for example, are nondegradable under conventional anaerobic digestion conditions even though a consortium of the microorganisms capable of degrading other biomass feeds is present.

It is generally thought that for most types of organic wastes and cellulosic biomass feedstocks a viable inoculum can be developed from any environment in which anaerobic methanogenic decomposition of a mixture of organic compounds is occurring naturally. Such environments include, for example, anaerobic sewage digesters, anaerobic lake sediments, or animal feces. Starting with such sources, an inoculum adapted to the feed under study can usually be developed within a reasonable period of time, that is, 2–4 months.

It would be logical, however, to expect that the biodegradation potential of an inoculum would improve after an extended period of adaptation to a feed. Such improvement was not observed, however, when using a kelp-adapted inoculum, for example.[1] Because kelp contains organic materials unique to marine plants (for example, mannitol, algin, fucoidin, and laminarin) and has a high salt content, it was presumed that a marine inoculum derived from anaerobic sediments containing decomposing kelp might be capable of greater decomposition of kelp than a sewage sludge inoculum. Several marine inocula, however, did not demonstrate greater kelp degradation than that observed with a sludge inoculum. There is a need to further evaluate inocula from different sources and their effects on the anaerobic digestion of biomass.

10.2.2 Effects of Reactor Size, Configuration, and Operating Conditions

Reactor size, configuration, and operating conditions can have a significant impact on the time required to start or recover dormant digesters. Large reactors require development of large inocula. Very large sources of inocula, especially those adapted to particular biomass feedstocks, are, however, frequently unavailable. Consequently, the inoculum must be grown by the steady addition of feedstock to a primary culture until the microbial population is of sufficient size to start the digester. Depending upon the relative size of the primary culture, it could take several months to start a large reactor.

The rate of digester start-up can best be estimated through smaller-scale studies using the primary culture source and reactor configuration intended for scale-up. Reactors operated at mesophilic temperatures are frequently more stable during start-up than thermophilically operated digesters.[2] Available cultures must be adapted to the elevated temperatures. Since culture stability is decreased under thermophilic conditions, special care is needed to control start-up conditions in order to assure stable operation.

Gradual increases in the loading rate are required in continuously stirred tank reactors (CSTR) to promote development of the methane-producing bacterial population for utilization of volatile acids at their rate of production. If the loading rate is increased too rapidly, the acid-forming bacteria produce volatile acids at a faster rate than they can be utilized. These volatile acids reduce culture alkalinity and, thus, promote digester imbalance through lower pH levels, which can be toxic to the methane-producing bacteria. Addition of sodium bicarbonate to a digester during start-up can reportedly decrease the time required for stable operation.[3] The buffer stabilizes digester pH fluctuations and aids in maintaining a balance between acid- and methane-forming bacteria.

Attached-film reactors such as anaerobic filter (AFR) or fluidized-bed (FBR) reactors require the development of a large population of micro-organisms on an inert, nonbiodegradable support medium. The start-up times, which can be relatively long, are determined by the kinetics of microorganism attachment and biofilm development under particular medium composition and fluid flow rate conditions within the reactor.

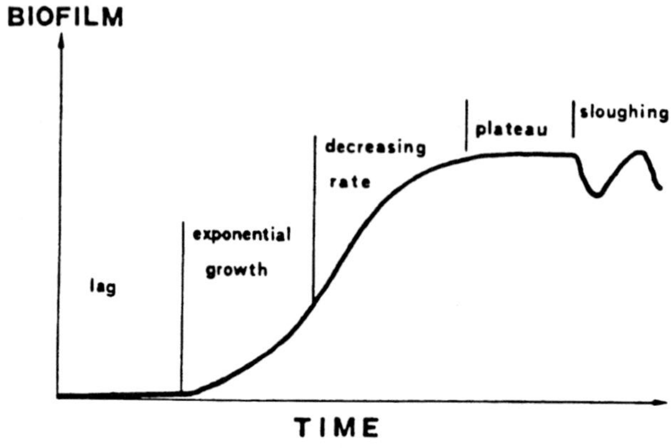

FIG. 10.1. Five stages of biofilm development.[2]

Bryers[4] described biofilm development in terms of the five stages illustrated in Fig. 10.1. The duration of the lag phase can vary with operating conditions such as inoculum size, fluid flow rates, and amount of recycled microorganisms. Although the type of media can affect start-up times, the time required for development of a suitable culture on bare support media in these reactors can be several months. Switzenbaum and Jewell,[5] for example, reported that up to 9 months were required to start an expanded-bed reactor. However, once established, the microbial population is quite stable. Such digesters can quickly be restarted, even after several months of dormancy.[6]

Improved methods for optimizing digester start-up and recovery in biomass digestion are needed and represent an important research requirement. Existing methods are not generally uniform and are dependent upon a host of variables, including inoculum composition, reactor configuration, and digester operating procedures. The factors that contribute to reliable and rapid digester recovery will have a significant impact on the feasibility of large-scale biomass anaerobic digestion processes.

10.3 OPERATION

10.3.1 Temperature

The temperature at which anaerobic digestion occurs can significantly affect the conversion, kinetics, stability, effluent quality, and net energy of the biological conversion process. Biological methanogenesis has been reported at temperatures ranging from 4 °C (lake sediments) to 60 °C (thermophilic dairy manure digesters). Thermophilic anaerobes capable of growth at temperatures as high as 97 °C have been isolated from naturally occurring environments such as hot springs.[7]

Although microorganisms with potential importance to the anaerobic digestion process can exist under a wide range of temperature conditions, most studies of practical process applications have been performed under ambient (20–25 °C), mesophilic (35 °C), or thermophilic (55 °C) temperature conditions. Studies have generally demonstrated that higher temperatures promote increased conversion rates but also decrease process stability.

10.3.1.1 Effects on Performance

It has generally been observed that higher temperatures promote higher reaction rates during anaerobic digestion, thus permitting lower hydraulic retention times and higher loading rates without reduction in conversion efficiency. In the range of 42–50 °C, reaction rates decrease below those observed in the mesophilic range and then increase as the thermophilic temperature range is reached. Gujer and Zehnder[8] described studies that demonstrated the extreme sensitivity of certain methanogenic bacteria to temperatures above the mesophilic range. Most digesters are operated at 35 °C because of high sensitivity of bacteria to temperature fluctuations in the thermophilic range.

Most of the experimental work evaluating the effects of temperature on anaerobic digestion has been performed with sewage sludge and manure. Comparing the performance of sludge digesters operated at temperatures ranging from 15 to 35 °C, O'Rourke,[9] for example, concluded that effective and efficient anaerobic treatment could be accomplished at reduced temperatures. At 25 °C, conversion was 90% of that observed at 35 °C. The performance at 15 °C was, however, drastically reduced and effective performance could not be achieved even after a 60-day retention time.

Higher temperatures generally increase the efficiency of organic solids destruction, improve dewatering characteristics of effluent solids, and increase the rates of destruction of pathogenic organisms.[10,11] The effects of higher temperatures on effluent quality are greater than those of mesophilic digesters at shorter retention times. As these retention times increase, the difference in the quality of the effluent from digesters operated in thermophilic and mesophilic ranges decreases. This concept was demonstrated by O'Rourke[9] and is illustrated for lower temperatures

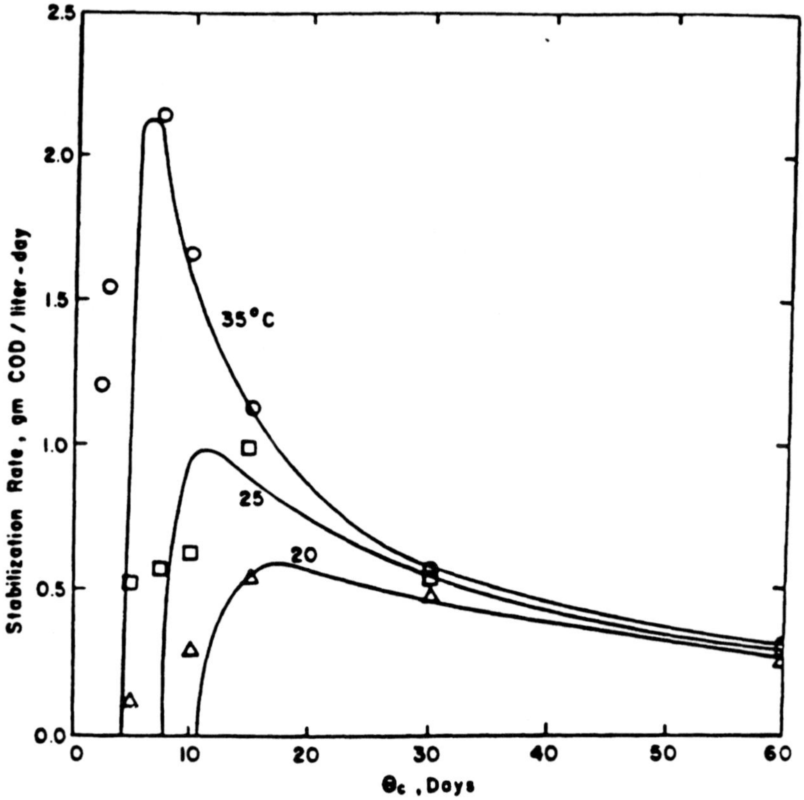

FIG. 10.2. Waste stabilization rate as a function of retention time and temperature for primary municipal sludge with S_T^0 = 28·4 g COD/liter and S_d^0 = 19·6 g COD/liter.[9]

in Fig. 10.2. Comparing sludge digesters at 20, 25, and 35 °C, the most rapid waste stabilization rate was observed at the highest temperature at short retention times. As the retention time increased to 60 days, however, very little difference was observed in the rate of waste stabilization among the digesters operated at each of the three temperatures.

A full-scale 8500-m^3 thermophilic digester was operated on a mixture of 90% activated sludge and 10% primary sludge at 54 °C with about an 11·3-day retention time.[12] The effluent quality was comparable to that obtained from mesophilic digestion, but the methane yield and production rates were higher than those obtained from a mesophilic digester.

The increased energy requirement, 120% over mesophilic, was believed to be compensated for by the increased gas production. Other studies using biomass feeds, on the other hand, have demonstrated inconsistent results when comparing the performance of thermophilic and mesophilic digesters. Some studies, for example, have demonstrated lower methane yields when biomass feeds were digested at higher temperatures than at lower temperatures.[9] Such findings were observed with Kentucky blue-grass and giant brown kelp. Higher methane yields were, on the other hand, demonstrated during the thermophilic digestion of coastal bermuda grass and Minnesota reed sedge peat.

10.3.1.2 Effects on Stability

Digester stability is critical to the feasibility of large-scale biomass digestion. The temperature of anaerobic digestion can markedly affect the stability of the process. Even fluctuations in the temperature within relatively narrow ranges under thermophilic conditions can have a significant impact on digester stability. Speece,[13] for example, reported on studies that showed that thermophilic digesters could only tolerate temperature changes of $0 \pm 0{\cdot}80\,°C$ and that the volatile acids concentrations were always more than twice as high as those in mesophilic digesters. Garber *et al.*[14] found that the volatile acids content of well-operated thermophilic sewage sludge digesters was six times greater than that of a mesophilic unit. They also observed that thermophilic organisms were more sensitive to rapid changes in temperature than mesophilic organisms.

10.3.1.3 Effects on Energy Requirements

Thermophilic digesters have higher heating requirements than digesters operated at lower temperatures. The significance of the additional required heat to the overall net energy efficiency of the digester depends upon the feed solids concentration and upon the relative methane production rates. Ghosh[15] showed, for example, that the net energy production of anaerobic digestion becomes negative for mesophilic and thermophilic digestion at solids concentrations below 2% and 4%, respectively. At loading rates of up to $8{\cdot}0\,kg\,VS/m^3$-day, heat losses, as percent of the product methane, could account for up to 20% and 50% in mesophilic and thermophilic digesters, respectively.

Important questions regarding the relative value of thermophilic digestion remain. As the solids concentrations of the feeds increase, as with many biomass species, the relative energy requirement for heating the feed to thermophilic temperatures decreases. However, several thermophilic digestion studies performed under more-or-less conventional conditions indicate decreased stability over mesophilic digestion. It has been reported that thermophilic digestion stability can be improved when preceded by mesophilic digestion.[16] Thermophilic digestion should be evaluated under conditions that otherwise promote digester stability, such as long solids and microorganism retention times.

At low solids concentrations, feed heating requirements can have a significant impact on the overall energy efficiency of the process. Energy-efficient techniques for effectively digesting biomass with a low solids (high water) content are needed. One method that should be studied further is to separate the water from the solid biomass components and digest each separately. The solid component could be digested at a higher temperature while the liquid component is digested at a lower temperature. Under such conditions, methods for improving the digestion rates at lower temperatures are needed. One method that should be investigated is that of developing inocula that can achieve higher rates of digestion at lower temperatures.

10.3.2 Pressure

Anaerobic digestion processes are generally performed under relatively low positive pressure. Maintaining a slight positive pressure of up to about 20·7 kPa (gauge) reduces the probability of reactor contamination by the introduction of air into the system. In large digesters, the microbial populations at the bottom of the reactors can be exposed to very high hydraulic pressures without documented effects on performance. In marine and ocean sediment degradation, Ort[17] demonstrated that relatively pure methane could be produced by subjecting conventional digesters to pressures of from 202·7 to 506·6 kPa (gauge). Dynatech Corporation operated anaerobic digesters at pressures as high as 3·1 MPa (gauge).[18] Such pressurized anaerobic digestion increased carbon dioxide availability as a hydrogen acceptor for biological methane gas formation.

Other studies, by contrast, have demonstrated improved performance under reduced pressure, or vacuum, conditions. Hashimoto,[19] for example, showed that the operation of a thermophilic beef waste digester under a vacuum of 15-inches water increased the methane production rate by as much as 5%. If a significant amount of digester gas is trapped in the digester culture, then decreased pressures can promote gaseous release and increased methane recoveries. On the other hand, operating a digester under low-pressure conditions will reduce the carbon dioxide pressure, which can, in turn, affect pH and alkalinity. Other studies suggested that reduced pressure increases the rate of conversion of acetate and methane.[20]

Anaerobic digestion can be performed at both higher and lower pressures with benefits that are characteristic of each condition. Further work is needed, however, to evaluate the relative benefits of pressure on the operation and performance of large-scale biomass digesters.

10.3.3 Mixing

Prior to 1950, most anaerobic digesters did not employ mechanical mixing.[21] The effect of nonmixing in sewage sludge digestion was that a floating scum layer formed at the surface of the culture, as illustrated in Fig. 10.3. Mixing was implemented as a means of disrupting this scum layer and thereby promoting greater digester homogeneity. Although the effects of mixing on anaerobic digestion are not well documented, it is believed by some that mixing also promotes improved digester performance by enhancing contact between organisms and substrates. The

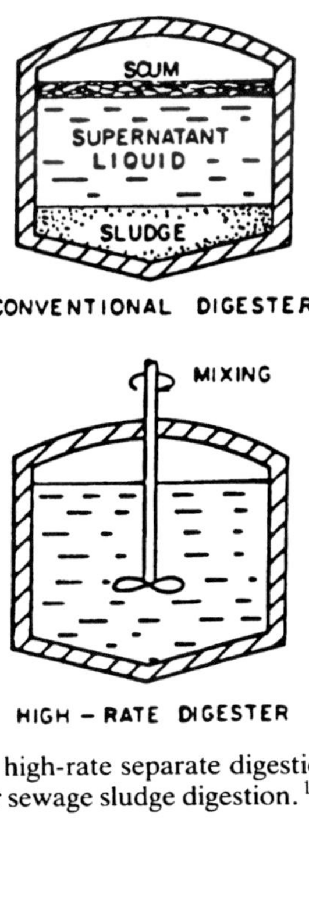

FIG. 10.3. Conventional and high-rate separate digestion systems commonly used for sewage sludge digestion.[14]

opposing views contend that mixing could destroy organisms or obligate interspecies associations and thus decrease digester performance. Furthermore, effects of mixing could depend on digester size and configuration, feed composition and particle size, method of mixing, and a number of other factors.

Studies with certain biomass substrates, such as kelp, have demonstrated that methane production rates are higher in two types of unmixed digester (baffle flow and upflow solids reactors), without attendant scum formation, than in continuously stirred tank reactors operated under similar conditions.[22] Studies with kelp that compared intermittent and continuous mixing showed lower methane yields with the latter technique.[23] It was hypothesized that the low specific growth rates associated with long digester residence times did not limit feed conversion efficiencies and substrate transport. Jewell *et al.*[24] also observed good fermentation of high-solids-containing agricultural residues in unmixed reactors. It should be noted that even mechanically unmixed reactors experience some culture agitation due to gas production.

Mixing requirements are related to the viscosity of the culture and to many other factors, such as density, particle size and shape, and water content. Achievement of reactor homogeneity with viscous or high-solids-content feeds is a very difficult task. Conventional mixing procedures are widely found ineffective in attaining homogeneity of the reactor contents. Dead zones occur unless the mixing equipment, reactor configuration, and energy input can all be harmoniously utilized to achieve the desired goal.[25] Homogeneously mixed digesters contain substrate solids, bacteria, and liquid dispersed evenly throughout the digester, and also in the effluent. In such reactors, the hydraulic, solids, and microorganism retention times are the same. Consequently, achieving long solids and microorganism retention times requires that the hydraulic retention time, and thus the reactor size, be large. The significance of this observation will be discussed below in more detail.

Ultimate digestion requires that the substrate come in contact with the bacteria or enzyme required for degradation. Dague[26] indicated that adequate contact between bacteria and the digesting substrate, through mixing, was required before complete stabilization could occur. It should be noted, however, that cellulose, which is contained in biomass, can be degraded by extracellular enzymes and that direct contact with bacteria is not essential to degradation.

Finney and Evans[27] proposed that phase transfer of gaseous products was rate limiting and, further, that the product gases inhibited the activities of methanogenic organisms. They suggested that at very high substrate concentrations, gas bubbles may surround bacteria and interfere with substrate diffusion into intracellular spaces. By employing vigorous agitation, the authors were able to increase the methane formation rate as much as six-fold. This increase was attributed to improved gas transfer characteristics caused by mixing.

Hashimoto[19] showed that continuous mixing of thermophilic beef waste digesters resulted in slightly higher methane production rates than when mixing 2 h/d. Since the anaerobic digestion process is carried out by diverse groups of bacteria, mixing can disrupt microbial consortia that could form a focus of biochemical activity. Further work is needed, however, to delineate the effects of mixing on mass transfer and microbial activity.

There are three principal methods of digester mixing: mechanical agitation, digester culture recirculation, and gas recirculation. Each of these methods was discussed by Stafford *et al.*[25] Of the mechanical agitation methods, the paddle or arm mixer, usually consisting of one or more blades attached to a rotating shaft, is probably the oldest. Using this method, it is difficult to achieve sufficient mixing to stop scum formation. Propeller-type mixers are best suited for low- to medium-solids-content feeds and are not suitable for feeds with higher solids content.

Culture recirculation can effectively reduce scum formation when properly utilized. This technique can be performed with external pumps and piping and affords the opportunity to introduce, for example, digester heating procedures. Gas recirculation also offers an effective method for digester mixing. Digester purging with certain gases, such as a mixture of hydrogen and carbon dioxide, can increase methane production but has no effect on conversion of nongaseous feeds.[28] Studies with draft tubes, multiple sequentially discharged lances, and floor-mounted diffuser boxes used for digester mixing showed that, when properly designed, each technique could maintain uniform digester contents.[29]

10.3.4 Feeding Frequency

The effects of feeding frequency on digester performance have not been conclusively determined. In practical applications, feeding is typically intermittent, about 1 to 3 times per day, rather than continuous and is based upon such rationale as work scheduling efficiency. From a biological point of view, as discussed by Speece,[13] digesters should be able to process more substrate when fed continuously rather than intermittently since, with the latter technique, substrate becomes alternately excessive and limiting. Because of the slow growth rates of the bacteria involved, however, the overall methane fermentation process reportedly exhibits continuous culture kinetics.[30]

Fannin *et al.*[31] demonstrated that significant variations in the methane content occur in digester gas produced at various intervals after once-per-day feeding. They also noted that digesters operated at the higher loading rates showed greater methane content variation than digesters operated at the lower loading rates. The methane content remained relatively stable in frequently fed digesters but fluctuated substantially in the once-per-day fed digester over a 24-hour period.

These data indicate that frequently fed digesters are more stable than once-per-day fed digesters, especially at higher loading rates. There remain, however, significant practical questions regarding the application of continuous feeding to large-scale digesters. For example, although the studies by Fannin *et al.*[31] showed significant diurnal fluctuations in methane content with once-per-day feeding, the actual methane yield was essentially the same as that observed in the frequently fed digester. In completely mixed reactors a portion of the material added during each feeding is wasted in the effluent. In frequently fed digesters, the effluent will contain portions of unreacted feed having very short residence times. The effects on digester performance of this feed short-circuiting need further documentation. Provided the digesting substrate maintains sufficient buffering capacity to keep the pH above toxic levels, little may be gained in terms of methane yields by increasing the frequency of digester feeding.

10.4 PROCESS STABILITY AND CONTROL

Large-scale implementation of anaerobic digestion processes for production of methane from biomass requires predictable performance. Centralized digestion facilities carry with them greater overall economic risks associated with operational failures due to equipment breakdowns and operational errors than more decentralized operations. The risks of economic losses due to digester failure can only be estimated if some understanding of process stability is developed. This understanding is based upon determination of the potential causes of digester imbalance, procedures to measure potential imbalance, and methods for restoring and maintaining stability.

10.4.1 Causes of Imbalance

Anaerobic digestion imbalance can be caused by any factors that disrupt the metabolic activities of any group of the consortium of bacteria that perform the diverse functions of digestion. Disruption of groups of organisms that utilize the metabolic products of other organisms can cause an accumulation of these products that can ultimately produce conditions that become toxic to the digestion process. Exposure of unadapted microorganisms to toxic substances can cause significant disruption of performance and potential failure of anaerobic digestion. Such disruptions can be caused by shock loading, unstable temperatures, or by the introduction of inhibitory substances such as oxygen, metals, organics, or antibiotics.

The effects of shock loading and temperature variations on anaerobic digestion were discussed above. Any activity or substance that either washes out or inhibits the methanogenic bacteria is likely to cause digester destabilization. Methanogens are relatively slow-growing and require a significant amount of time to reproduce themselves to levels that enable them to utilize the fermentation products produced by faster-growing bacteria. Consequently, the time required for digester recovery after disruption of the methanogenic bacterial population can be significant. Studies by Speece,[13] however, demonstrated that most inhibitors show a reversible bacteriostatic effect on methanogens in the concentrations usually found in industrial waste streams.

Certain agricultural wastes can contain inhibitory concentrations of substances such as ammonia or antibiotics. High ammonia concentrations produced during the anaerobic digestion of piggery wastes exhibited inhibitory effects, especially at high pH levels.[32] Varel and Hashimoto[33] studied the effect of animal wastes containing several antibiotics. They observed that the degree of inhibition varied with different antibiotics and that, after an initial acclimation period, cultures adaptation occurred. Ammonia and antibiotic inhibition are unlikely to occur in biomass feeds. There is, however, a lack of information on the potential effects of several kinds of herbicides or pesticides that might be used with biomass crops.

10.4.2 Measures of Imbalance

Several parameters are traditionally used to determine the stability and performance of anaerobic digestion processes: These include gas production and composition, reduction in organic matter, pH, alkalinity, volatile acids, and other parameters such as oxidation–reduction potential, hydrogen production, and effluent ammonia concentrations.

10.4.2.1 Gas Production and Composition

The ultimate products of biomethanogenesis are methane and carbon dioxide. Although gas production is an indicator of microbial activity, methane content is probably one of the best indicators of overall digester performance. During stable digestion with a nonvariable feed, relatively stable gas production with a predictable methane content can be expected. When gas production or composition is erratic, there is an indication of digester instability. Low methane content is an initial indicator of the inhibition of methanogenic bacteria, as the amount of gas produced and its composition are relatively simple to measure.

10.4.2.2 Reduction in Organic Matter

Volatile solids (ash-free dry weight) are commonly used as a measurement of organic matter in determining the loading and conversion efficiency during anaerobic digestion. Additional measurements of organic matter — such as: chemical oxygen demand; carbon in the feed, effluent, and gas; and heating values of feed and effluent — are often also utilized for estimation of organic composition.

10.4.2.3 pH and Alkalinity

Digester pH, or hydrogen ion activity, is a measure of the acidic or basic nature of the culture. The optimum pH for anaerobic digestion is normally in the range of 7–8; pH levels that deviate significantly from this range can indicate potential toxicity and digester failure. Low pH levels, for example, can be a symptom of digester imbalance. As volatile acids concentrations increase, the pH in the digester decreases. At pH levels below 6·0, the acidic conditions produced can become toxic to methane bacteria. The pH of anaerobic digestion is a function of the carbon dioxide content of the gas, as well as the alkalinity.[34]

Alkalinity is a measure of the buffering or acid-neutralizing capacity of the digester. Since this buffering capacity can be provided by a wide range of substances, total alkalinity is a non-specific measure that is frequently interpreted as a determination of the carbonate, bicarbonate, and hydroxide content of a culture. At pH levels ranging from 6 to 8, the carbon dioxide–bicarbonate system is the major chemical system controlling alkalinity. Under these conditions, bicarbonate alkalinity is the major component of the total alkalinity and is represented by the following equilibrium equation:

$$H_2CO_3 \leftrightarrows H^+ + HCO_3^- \tag{10.1}$$

The dissociation of carbonic acid accounts for most of the alkalinity in anaerobic digestion.[35] Bicarbonate alkalinity, pH, and carbon dioxide pressure are interrelated in anaerobic digestion.[34] When the bicarbonate alkalinity falls below 500 mg/liter, with normal carbon dioxide production, the pH approaches 6·0. Thus, alkalinity destruction results in decreased pH levels.

Effective pH control requires that the alkalinity of the digester be increased. Maintenance of bicarbonate alkalinity above about 2500 mg/liter should be promoted to enhance digester stability. As digester alkalinity is depleted by volatile acids accumulation, for example, the pH in the digester also decreases. Since bicarbonate alkalinity is destroyed by increased volatile acids concentrations in destabilized digesters, excess alkalinity is needed when these acids are produced.[34] Decreases in bicarbonate alkalinity decrease digester stability.

Depending on the stability of the anaerobic digestion process and its potential for producing alkalinity, means for increasing alkalinity and pH may be needed. Attempts to increase pH without increasing alkalinity may be insufficient to recover an unstable digester. To increase alkalinity and control pH in laboratory-scale digesters, sodium bicarbonate is frequently added either in place of or in addition to sodium hydroxide. Excessive sodium concentrations, however, can be toxic to the digestion process and should be avoided.

Alkalinity may also be increased by adding lime, or it may be generated through protein degradation to ammonia, which can combine with the carbonic acid in solution to form an ammonium bicarbonate buffer. The specific method used for maintaining alkalinity and pH in digesters is affected by their relative impacts on process economics. Where possible, digesters should be operated under the conditions that promote alkalinity and stable pH, without external additives.

There is a need for additional research to establish digester operating procedures that promote alkalinity. Furthermore, there is a need for developing methods for accurately measuring alkalinity in the presence of high solids concentrations such as those observed in some biomass digestion processes. Measurement in anaerobic digesters can be affected by such interfering substances as soaps, oily matter, suspended solids, or precipitates that can contribute to imprecise titration endpoints.[36]

10.4.2.4 Volatile Acids

Volatile acids represent a critical parameter in the operation and control of anaerobic digestion. Volatile acids fermentation products such as acetic, propionic, and butyric acids are formed by 'acid-forming' bacteria through the fermentation of hydrolytic products such as soluble carbohydrates, long-chain fatty acids, and amino acids. Volatile acids are also produced from fats and through certain reactions with hydrogen. For example, acetate can be formed from carbon dioxide and hydrogen; propionate may be formed from acetate, carbon dioxide, and hydrogen. The most important volatile acid with respect to methane formation during anaerobic digestion is acetic acid. Acetic acid occurs primarily in an extracellular pool, has a rapid turnover rate,[37] and accounts for most of the methane produced during anaerobic digestion. Jeris and McCarty[38] and Weng and Jeris[39] showed that 80% or more of the methane can be produced from acetate during the degradation of the amino acids glutamic acid and isoleucine.

The concentration of these volatile acids in anaerobic digesters is determined by their rate of production and their rate of removal. When their rate of production exceeds their rate of removal, volatile acids concentrations increase. In a healthy digester, there is a balance between the amount of volatile acids produced and the amount removed through conversion by 'methane-forming' bacteria. Metabolism of certain fatty acids is strongly inhibited in the presence of trace concentrations of hydrogen. This explains, in part, the difficulty of restoring stability to an inhibited digester. It further supports the utility of monitoring individual volatile acids as a measure of digester stability. Also, in an unstable digester the propionate-utilizing population must develop to accommodate its sudden increased concentration.

Any factors that selectively stimulate acids production or inhibit the slow-growing methane-forming bacteria are likely to cause an accumulation of volatile acids in the digester. Such factors include rapid increases in the organic loading rate, the introduction of inhibitory substances into the digester, or extreme changes in the digester temperature. The effects of rapid loading-rate increases on kelp digestion, for example, are illustrated in Fig. 10.4.[31] Using a single kelp lot, the loading rate in this CSTR digester was increased to 4·8 kg VS/m^3-day. After about 30 days,

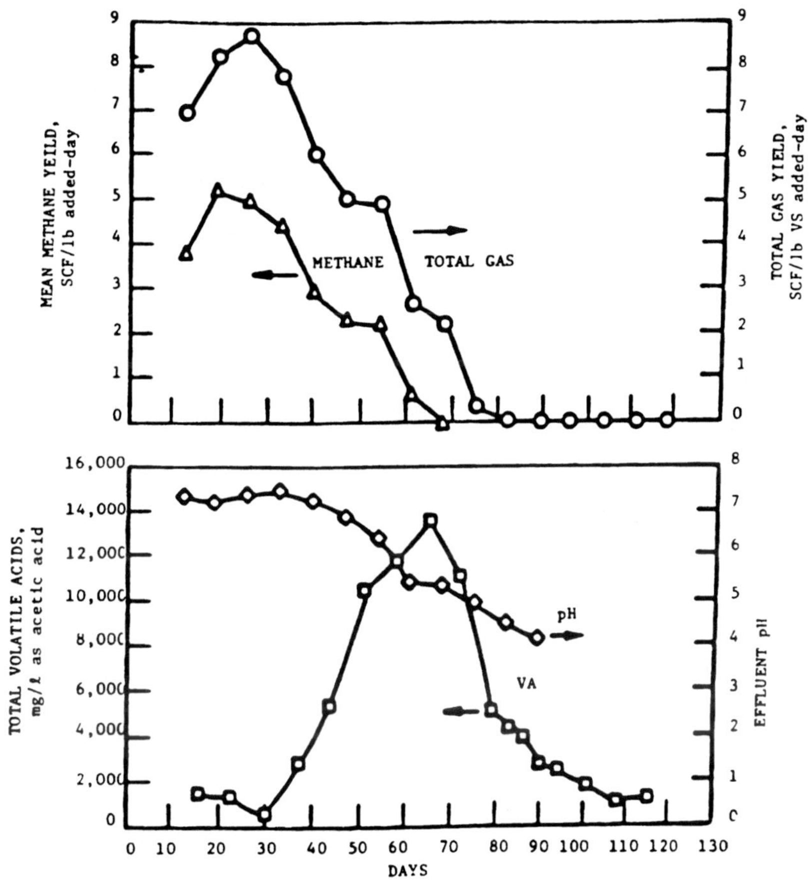

FIG. 10.4. Performance curves for acid-phase digester receiving kelp lot 50 at 0·30 lb VS/ft^3-day loading without pH adjustment (STR; culture volume 10 liters; HRT 13 days; 35 °C).

the volatile acids concentration began to increase rapidly and the methane yield fell. As the volatile acids concentration increased to a maximum value of 13 500 mg/liter as acetic, the pH dropped to near 5 and gas production dropped to nil. The pH continued to decline to 4 and there was no evidence of fermentation activity. As the volatile acids concentration fell to the levels observed before digester destabilization, recovery did not occur.

Changes in digester operating conditions should be made gradually to prevent the accumulation of excess volatile acids. Di San Marzano *et al.*[40]

recommended that the loading rate in CSTRs should be no more than doubled at maximum 20% increments over a 14-day period and that each change in operating conditions should be followed by a steady-state period, in which stable performance is demonstrated for a specified time, before additional changes. If changes occur too rapidly, propionate can, for example, accumulate and inhibit the acid-forming bacteria.[41]

Separating the acid- and methane-producing bacteria in two-phase reactor systems can minimize the impact of high volatile acids concentrations on methanogens. In fact, the volatile acids concentrations can be decreased by recycling the effluent from the first phase back in the second phase.

Discrepancies remain regarding the effects of volatile acids and pH on digester stability. High volatile acids concentrations are reported by numerous investigators to be a symptom rather than a cause of decreased digester stability. In fact, sudden increases in the concentrations of such volatile acids as acetic and butyric reportedly stimulate, rather than inhibit, anaerobic digestion.[42] Schwartz and Keller,[43] on the other hand, found that the conversion of glucose to acetate and acetic acid by *Clostridium thermoaceticum* was inhibited more by free acetic acid than by either acetate ions or low pH.

The production of volatile acids during anaerobic digestion gradually shifts the thermodynamics of the microorganism culture toward an environment where appropriate levels of obligate proton-reducing bacteria are also maintained.[40] This promotes greater digester stability during shock caused by increased volatile acids concentrations. Increasing concentrations of magnesium acetate up to 4000 mg/liter can stimulate gas production in laboratory-scale sewage sludge digesters, with a maximum stimulation achieved at a concentration of 1000 mg/liter.[44] Good gas yields were observed when the ratio of propionic and butyric acids to acetic acid was less than 80:1. Gas production was inhibited at higher ratios.

More work is needed on the development of techniques to remove excessive volatile acids concentrations from anaerobic digesters. Of several methods that have been proposed, the techniques that promote recovery of the potential methane from the volatile acids appear most promising. Such procedures may involve volatile acids removal from the digester via dialysis, followed by digestion in a second-stage digester. A second procedure involves the application of two-phase digestion with culture recycle. Using such a technique, the effluent from the digester containing high volatile acids concentrations is fed to a second-phase attached-film digester. The volatile acids are removed in the second phase and the effluent is returned to the first phase. Such techniques are under evaluation for application to marine biomass feeds.[6] A well-operated digester should not require techniques to control the volatile acids concentration unless it is overloaded or is subjected to otherwise stressful conditions.

10.4.2.5 Other Parameters

Anaerobic digestion involves a complex chain of biochemical reactions that are determined by the thermodynamic state of the system. Oxidation–reduction reactions, as described by Diriasian *et al.*,[45] enable the micro-organisms to satisfy their immediate needs for energy as well as to store energy for future utilization; they can be described by the following equation:

$$E_h = E_0 + (RT/nF)\{\ln[(\text{Oxid})/(\text{Red})]\} \qquad (10.2)$$

where E_h = the potential of the system referred to the normal hydrogen electrode (volts), E_0 = a specific constant for the system being measured (volts), R = universal gas constant, T = absolute temperature, F = Faraday constant, n = number of electrons transferred in the oxidation–reduction reaction.

When the ratio of oxidant to reductant increases, the potential becomes more positive; and when the ratio decreases, the potential becomes more negative. Thus, oxygen, an oxidizing agent, imparts a high positive potential; while hydrogen, a reducing agent, demonstrates a high negative potential.

Methane-producing bacteria are obligate anaerobes. Metabolic functions are performed in a highly reduced environment having a highly negative potential. Thus, the oxidation–reduction potential of an anaerobic digestion culture should be low. Diriasian *et al.*[45] reported that optimum digestion appeared to be a function of healthy methane-producing flora that seemed to thrive best at an E_c level between −520 and −530 mV (E_c was used in preference to E_h to describe measured rather than calculated values). Zehnder[46] reported inactivation of methanogens at −220 mV. Schwartz and Keller[43] showed that the growth of *Clostridium thermoaceticum* did not occur when the redox potential exceeded −300 and −360 mV at pH values of 6 and 7, respectively.

Although the importance of oxidation–reduction potential to anaerobic digestion is widely recognized, its utilization as an operation control parameter is not well documented. It has potential application as a rapid automated process control parameter. Further work is, however, needed to document the utility of oxidation–reduction potential as a measurement of digester stability.

The conversion of alcohols, propionate, and longer-chain fatty acids to acetate and hydrogen is performed by obligate proton-reducing microorganisms. These organisms must exist in a symbiotic relationship with hydrogen-utilizing methanogens that can convert acetate and hydrogen to methane.[47] As noted above, it is important that this hydrogen be removed to prevent its accumulation since high hydrogen partial pressures create thermodynamic conditions that are unfavorable for this conversion process. The effect of hydrogen partial pressure on the free energy of this conversion was discussed by McCarty.[21] According to McInerney and Bryant,[48] the addition of hydrogen inhibits acetate degradation by certain methanogenic bacteria. During the performance of a healthy digester, hydrogen does not accumulate but is converted to methane. Boone[49] found that the addition of hydrogen to an animal-waste digester resulted in an increase in the acetate pool.

More research is needed on the significance of hydrogen in anaerobic digestion. Methods to reduce the accumulation of hydrogen in digesting culture are needed. Sensitive methods for measuring very low concentrations of hydrogen should be developed to enable the utilization of this parameter for routine evaluation of digester operating performance.

10.4.3 Restoring and Maintaining Stability

Restoration of digester stability requires that the underlying cause of destabilization be identified and that corrective measures be implemented. In addition, further steps may be required to correct symptoms created by the imbalance. For example, shock loading can result in an accumulation of volatile acids caused by a relative imbalance created in the concentration of acid- and methane-producing bacteria. One way to promote digester restabilization is to reduce the loading rate and promote an increase in the methanogen population through utilization of accumulated volatile acids. If, however, the alkalinity of the digester is low, the pH and alkalinity may require adjustment to non-toxic levels before digester recovery can occur.

Long solids retention times can reduce toxicity problems and promote digester stability without significantly sacrificing process efficiency.[50]

Provided the methane-producing bacterial population can be retained in the digester, even under high-loading conditions, increased digester stability can be expected. Consequently, reactor designs, such as those discussed in Chapter 9, that promote long solids and microorganisms retention times can be expected to promote stable digestion.

Attached film reactors have been reported to be extremely stable during stressful conditions of anaerobic digestion. Studies in an anaerobic attached film expanded-bed reactor, for example, demonstrated that variations in temperature, flow rate, or concentration of synthetic wastewater produced no significant shock effects.[51] This high degree of stability was attributed to exceptionally high bacterial cell concentrations in the reactor and to a microorganism retention time in excess of 1 year. Studies with a downflow anaerobic filter reactor operating on chemical industry wastes that received 8 times the normal loading rate for a 24-hour period recovered within 12 to 48 hours after overloading was stopped.[52] Anaerobic filter reactors have been reported to be more resistant to temperature- or loading-induced shocks and recover more rapidly than upflow sludge blanket reactors.[53]

10.5 SUMMARY

Digester start-up is affected by the characteristics of the inoculum, biomass feed, reactor configuration and size, and operating conditions. Improved methods for optimizing digester start-up and recovery in biomass digestion are needed and represent an important research need. Existing methods are not generally uniform and are dependent upon several variables, including inoculum composition, reactor configuration, and digester operating procedures. Factors that contribute to reliable and rapid digester recovery will have a significant impact on the feasibility of large-scale biomass anaerobic digestion processes. The relevance of developing specialized inocula in anaerobic digestion has not been conclusively demonstrated. There is a need to further evaluate adapted inocula and their effects on the anaerobic digestion of biomass.

The temperature at which anaerobic digestion occurs can significantly affect the conversion, kinetics, stability, effluent quality, and net energy of the biological conversion process. Important questions regarding the relative value of thermophilic digestion remain. The energy requirement for heating the feed to thermophilic temperatures decreases inversely with the solids concentration of the feedstock. However, several thermophilic digestion studies, performed under more-or-less conventional conditions, indicate less stability than in mesophilic digestion. Thermophilic digestion should be evaluated under conditions that otherwise promote digester stability, such as long solids and microorganism retention times.

At low solids concentrations, feed heating requirements can have a significant impact on the overall energy efficiency of the process. Energy-efficient techniques for effectively digesting biomass with a low solids (high water) content are needed. One method that should be studied further is to separate the water from the solid biomass components and digest each separately. The solid component could be digested at a higher temperature while the liquid component is digested at a lower temperature.

The effects of pressure on anaerobic digestion are not conclusively determined. There are conflicting data regarding the relative benefits of high and low pressures. Further work is needed before the benefits of relative pressure levels can be determined and applied to large-scale biomass digesters.

A major reason for introducing mixing into anaerobic digestion processes was to reduce scum formation. This problem may be less significant with biomass feeds. The relative importance of mixing for improving digester performance by enhancing contact between microorganisms and substrates has not been conclusively documented. Further work is needed to delineate the effects of mixing on mass transfer, microbial activity, and materials handling in digesters.

The feasibility of large-scale biomass digestion is highly dependent upon the stability of the process. Methods are needed to promote and assure stability as well as for the rapid recovery of destabilized digesters. There is a need for additional research to establish digester operating procedures that promote alkalinity. Furthermore, there is a need for developing methods for accurately measuring alkalinity in the presence of high solids concentrations such as those observed in some biomass digestion processes. More work is needed on the development of techniques to reduce excessive volatile acids concentrations in anaerobic digesters. Such procedures may, for example, involve volatile acids removal from the digester via dialysis, followed by digestion in a second-stage digester or the application of two-phase culture recycle. Using the latter technique, the volatile-acids-containing effluent from the first-phase digester are removed to a second-phase digester, and the effluent is returned to the first phase.

Although the importance of oxidation–reduction potential to anaerobic digestion is widely recognized, its utilization as an operation evaluation parameter is not well documented. It has potential application as a rapid automated process control parameter. Further work is needed to document the utility of routine measurement of the oxidation–reduction potential for process control in biomass digesters. More research is needed on the significance of hydrogen in anaerobic digestion. Sensitive methods for measuring very low concentrations of hydrogen should be developed to enable the utilization of this parameter for routine evaluation of digester operating performance.

REFERENCES

1. Chynoweth, D. P., 'Microbial Conversion of Biomass to Methane,' Paper presented at the Eighth Annual Energy Technology Conference and Exposition, Washington, DC, March 1981.
2. Rimkus, R. R., *et al.*, 'Full-Scale Thermophilic Digestion at the West-Southwest Sewage Treatment Works, Chicago, Illinois,' *Jour. Wat. Poll. Con. Fed.* **54**:1447, 1982.
3. Rupert, H., 'Sodium Bicarbonate Smoothes Digester Start-Up,' *Water Sew. Works* **125**:58, 1978.
4. Bryers, J. D., 'Processes Involved in Early Biofilm Formation,' in *Proceedings: First International Conference on Fixed-Film Biological Processes*, Wu, Y. C., Smith, E. D., Miller, R. D. and Patken, E. J. O., Eds. Pittsburgh: University of Pittsburgh, 1982.
5. Switzenbaum, M. S. and Jewell, W. J., 'Anaerobic Attached-Film Expanded-Bed Reactor Treatment,' *J. Water Poll. Cont. Fed.* **52**:1953–1965, 1980.
6. Fannin, K. F., Srivastava, V. J., Mensinger, J. D. and Chynoweth, D. P., 'Marine Biomass Program: Anaerobic Digestion Systems Development and Stability Study,' *IGT Project 30558 Final Rep.*, Chicago, June 1983.
7. Ljungdahl, L. G., *et al.*, 'Some Aspects of Thermophilic and Extreme Thermophilic Anaerobic Microorganisms,' *Trends Biol. Ferment.* **18**:391, 1981.
8. Gujer, W. and Zehnder, A. J. B., 'Conversion Processes in Anaerobic Digestion,' *Water Science and Technol.* **15**:127–167, 1983.
9. O'Rourke, J. T., 'Kinetics of Anaerobic Waste Treatment at Reduced Temperatures,' Doctoral Dissertation, Stanford University, 1968.
10. Buhr, H. O. and Andrews, J. F., 'The Thermophilic Anaerobic Digestion Process,' *Water Res.* **11**:129–143, 1977.
11. Garber, W. F., 'Operating Experience With Thermophilic Anaerobic Digestion,' *J. Water Poll. Cont. Fed.* **54**:1170–1175, 1982.
12. Ghosh, S., Klass, D. L., Christopher, R. W. and Edwards, V. H., 'Thermophilic Biogasification of Biomass,' Paper presented at 7th Energy Technology Conference and Exposition, Washington, DC, March 1980.

13. Speece, R. E., Environmental Requirements for Anaerobic Digestion of Biomass, *Advances in Solar Energy, An Annual Review of Research and Development* **2**:51–123, 1985.
14. Garber, W. F. *et al.*, 'Thermophilic Digestion at the Hyperion Plant,' *J. Water Poll. Cont. Fed.* **47**:950, 1975.
15. Ghosh, S., 'Net Energy Production in Anaerobic Digestion,' pp. 253–273 in Symposium Papers — *Energy From Biomass and Wastes V*. Chicago: Institute of Gas Technology, 1981.
16. Hashimoto, A. G., 'Methane Production and Effluent Quality From Fermentation of Beef Cattle Manure and Molasses,' *Biotechnol. Bioeng. Symp.* **11**:481, 1981.
17. Ort, J. E., 'High Quality Methane Gas Through Modified Anaerobic Digestion,' US Patent 3 981 000 (21 September, 1976).
18. Wise, D. L., Cooney, C. L. and Augenstein, D. C., 'Biomethanation, Anaerobic Fermentation of CO_2, H_2, and CO to Methane,' *Biotechnol. Bioeng.* **20**:1153–1172, 1978.
19. Hashimoto, A. G., 'Effect of Mixing Duration and Vacuum on Methane Production Rate From Beef Cattle Waste,' *Biotechnol. Bioeng.* **24**:9, 1982.
20. Finney, C. D., Evans, R. S., II, and Finney, K. A., 'The Fast Production of Methane by Anaerobic Digestion,' US DOE Progress Report (24 May 1977–23 May 1978) No. C00-2900-16, Des Moines, Iowa: Natural Dynamics, June 1978.
21. McCarty, P. L., 'One Hundred Years of Anaerobic Treatment,' Paper presented at the Second International Conference on Anaerobic Digestion, Travemunde, Germany, 7 September 1981.
22. Fannin, K. F., Srivastava, V. J. and Chynoweth, D. P., 'Unconventional Anaerobic Digester Designs for Improving Methane Yields From Sea Kelp,' in Symposium Papers, *Energy From Biomass and Wastes VI*, sponsored by Chicago: Institute of Gas Technology, 1982.
23. Ghosh, S., Klass, D. L. and Chynoweth, D. P., 'Bioconversion of *Macrocystis pyrifera* to Methane,' *J. Chem. Technol. and Biotechnol.* **31**:791–807, 1981.
24. Jewell, W. J. *et al.*, 'Dry Fermentation of Agricultural Residues,' Final Report for Subcontract No. XB-9038-i, Solar Energy Research Institute, for US Department of Energy Contract No. EG-77-C-01-4042, December 1982.
25. Stafford, D. A., Hawkes, D. L. and Horton, R., *Methane Production From Waste Organic Matter*. Boca Raton, Fla: CRC Press, 1980.
26. Dague, R. R., 'Application of Digestion Theory to Digester Control,' *J. Water Poll. Cont. Fed.* **40**:2021–2032, 1968.
27. Finney, C. D. and Evans, R. S., II, 'Anaerobic Digestion: The Rate-Limiting Process and the Nature of Inhibition,' *Science*, **4219**:1088, 1975.
28. Chynoweth, D. P., Srivastava, V. J., Henry, M. P. and Tarman, P. B., 'Biothermal Gasification of Biomass,' p. 527 in *Symposium Papers: Energy From Biomass and Wastes IV*, D. L. Klass and J. W. Weatherly III, Eds. Chicago: Institute of Gas Technology, 1980.
29. Baumann, P. G. and Huibregtse, C. L., 'Evaluation and Comparison of Digester Gas Mixing Systems,' *Jour. Wat. Poll. Con. Fed.* **54**:1194, 1982.
30. Chynoweth, D. P., Ghosh, S. and Klass, D. L., 'Anaerobic Digestion of Kelp,' Chapter 18 in *Biomass Conversion Processes for Energy and Fuels*,

O. R. Zaborsky and S. Sofer, Eds. New York: Plenum Publishing Corp., 1981.

31. Fannin, K. F., Srivastava, V. J., Mensinger, J., Conrad, J. R. and Chynoweth, D. P., 'Marine Biomass Program: Anaerobic Digester Process Development,' *IGT Projects 65044 and 30547 Final Rep.* Chicago, 1982.
32. Van Velsen, A. F. M., 'Anaerobic Digestion of Piggery Waste. II. Startup,' *Neth. J. Agric. Sci.* (The Netherlands) **27**:142, 1979.
33. Varel, V. H. and Hashimoto, A. G., 'Methane Production by Fermentor Cultures Acclimated to Waste From Cattle Fed Monensin, Lasalocid, Salinomycin, or Avopartin,' *Appl. Environ. Microbiol.* **44**:1415, 1982.
34. McCarty, P. L., 'Anaerobic Processes,' Paper presented at the Birmingham Short Course on Design Aspects of Biological Treatment, Internat. Assoc. Water Poll. Res., Birmingham, England, 1974.
35. Capri, M. G. and Marais, G. R., 'pH Adjustment in Anaerobic Digestion,' *Water Res.* **9**:307, 1965.
36. *Standard Methods for the Examination of Water and Wastewater*, 15th edn Washington, DC: American Public Health Association, 1981, p. 254.
37. Smith, P. H. and Mah, R. A., 'Kinetics of Acetate Metabolism During Sludge Digestion,' *Appl. Microbiol.* **14**:368–371, 1966.
38. Jeris, J. S. and McCarty, P. L., 'The Biochemistry of Methane Fermentation Using ^{14}C Tracers,' *J. Water Poll. Cont. Fed.* **37**:178–192, 1965.
39. Weng, C. and Jeris,. J. S., 'Biochemical Mechanisms in the Methane Fermentation of Glutamic and Oleic Acids,' *Water Res.* **10**:9–18, 1976.
40. Asinari Di San Marzano, C.-M., *et al.*, 'Volatile Fatty Acids, an Important Parameter for the Control of the Reliability and the Productivities of Methane Anaerobic Digestion,' *Biomass* **1**:47, 1981.
41. McCarty, P. L., Jeris, J. S. and Murdoch, W., 'Individual Volatile Acids in Anaerobic Treatment,' *J. Water Pol. Cont. Fed.* **35**:1501–1516, 1963.
42. McCarty, P. L. and Brosseau, M. H., 'Effect of High Concentrations of Individual Volatile Acids on Anaerobic Digestion,' Paper presented at the 18th Annual Purdue Industrial Waste Conference, Lafayette, Indiana, 1963.
43. Schwartz, R. D. and Keller, F. A., Jr, 'Acetic Acid Production by *Clostridium thermoaceticum* in pH Controlled Batch Fermentations at Acidic pH,' *Appl. Environ. Microbiol.* **43**:1385, 1982.
44. Stafford, D. A., 'The Effects of Mixing and Volatile Fatty Acid Concentrations on Anaerobic Digester Performance,' *Biomass* **2**:43, 1982.
45. Diriasian, H. A., Molof, A. H. and Borchardt, J. A., 'Electrode Potentials Developed During Sludge Digestion,' *J. Water Poll. Cont. Fed.* **35**:424–439, 1963.
46. Zehnder, A. J. D., 'Ecology of Methane Fermentation,' Chapter 13, in *Microbiology of Water Pollution*, R. Mitchell, Ed. New York: Wiley Interscience, 1978.
47. Mah, R., 'Methanogenesis and Methanogenic Partnerships,' *Philosoph. Trans. Roy. Soc., London*, **297**:599–616, 1982.
48. McInerney, M. J. and Bryant, M. P., 'Anaerobic Degradation of Lactate by Syntrophic Associations of *Methanosarcina barkeri* and *Desulfovibrio* Species and Effect of H_2 on Acetate Degradation,' *Appl. Environ. Microbiol.* **41**:346–354, 1981.

49. Boone, D. R., 'Terminal Reactions in the Anaerobic Digestion of Animal Waste,' *Appl. Environ. Microbiol.* **43**:57, 1982.
50. Parkin, *et al.*, 'Modeling Toxicity in Methane Fermentation Systems,' *Jour. Environ. Eng. Div.* **108**:515, 1982.
51. Jewell, W. J. and Morris, J. W., 'Influence of Varying Temperature, Flowrate and Substrate Concentration on the Anaerobic Attached-Film Expanded-Bed Process,' *Proc. 36th Ind. Waste Conf. (1981), Purdue Univ.* Ann Arbor, Mich.: Ann Arbor Science Publishers, Inc., 1982.
52. Kennedy, K. J. and Van den Berg, L., 'Effects of Temperatures and Overloading on the Performance of Anaerobic Fixed-Film Reactors,' p. 678 in *Proc. 36th Ind. Waste Conf. (1981), Purdue Univ.* Ann Arbor, Mich.: Ann Arbor Science Publishers, Inc., 1982.
53. Van den Berg, L., *et al.*, 'Effect of Type of Waste on Performance of Anaerobic Fixed-Film and Upflow Sludge Bed Reactors,' p. 680 in *Proc. 36th Ind. Waste Conf. (1981), Purdue Univ.* Ann Arbor, Mich.: Ann Arbor Science Publishers, Inc., 1982.

11

Kinetics and Modeling

Y. R. CHEN
USDA Meat and Animal Research Center,
Clay Center, Nebraska, USA

and

VIPUL J. SRIVASTAVA
Institute of Gas Technology, Chicago, Illinois, USA

11.1 INTRODUCTION

An understanding of process kinetics is essential for rational reactor design; specifying operating conditions; and predicting methane production, system stability, and effluent quality. The full potential of the anaerobic process to produce methane cannot be achieved without formulating a rational basis for the process analysis and design. Since anaerobic digestion is a complex biological process involving decomposition by several major populations of microorganisms, the kinetic equations should be consistent with the underlying metabolic and biological principles governing the whole process. Also, the kinetics or the rate of reaction is different for different substrates, and for each substrate it is different at different operating conditions. This can be documented with a great degree of reliability by using the first-order kinetic model. Considering these variables, various kinetic equations are developed for each case and then these kinetic equations, along with equations developed from material balances of specific systems, can be used to predict the behavior of microorganisms with respect to changes in operating conditions and to provide guidelines for scale-up of the digestion systems.

11.2 GENERAL KINETIC EQUATIONS AND MODELS

Two basic relationships must be established when studying microbial kinetics: one relates microbial growth to substrate utilization, and the other relates microbial growth (or substrate utilization) to substrate concentration and microbial mass concentration.

The relationship between microbial growth and substrate utilization is formulated as

$$q = \mu/Y \tag{11.1}$$

where μ is the microbial specific growth rate, Y is the growth yield coefficient, and q is the specific rate of microbial substrate utilization.

The growth yield coefficient (Y) has been reported to decrease at low specific growth rates. The reduced growth yield coefficient at a low specific growth rate has been attributed to the existence of a maintenance energy requirement for cell growth.[1] The growth yield coefficient as a function of maintenance energy requirement and microbial specific growth rate has been expressed as[2,3]

$$Y = Y_G/[1 + (M/\mu)\,Y_G] \tag{11.2}$$

where Y_G is the true growth yield constant for the portion of the substrate that is converted to cells, and M is the maintenance energy requirement.

Different approaches to varying the growth yield coefficient have been attempted. A term for endogenous respiration (K_d) was incorporated into eqn (11.1):

$$q = (\mu + K_d)/Y_G \tag{11.3}$$

where Y_G is the growth yield constant, and K_d is the endogenous respiration rate. The concept of endogenous respiration has been adopted widely in describing biological waste treatment kinetics.[1] Kirsch and Sykes,[2] however, have shown that inclusion of a term for endogenous respiration (K_d) is mathematically equivalent to inclusion of maintenance energy (M), a term describing the decreasing growth yield coefficient at low specific cell growth rates.

Another relationship relates microbial specific growth rate (or specific substrate utilization rate) to substrate concentration and microbial mass concentration. Several models given below have been proposed to describe this relationship.

Early kinetic studies assumed that the growth of mixed cultures in complex organic wastes would be similar to the growth of a pure culture utilizing simple substrates. The Monod[4] model, which describes the kinetics of pure culture utilizing a single substrate, is represented by

$$\mu = (\bar{\mu}S)/(K_s + S) \tag{11.4}$$

where $\bar{\mu}$ is the maximum microbial specific growth rate, K_s is the half-velocity constant, and S is the limiting substrate concentration. Equation (11.4) has been used to study the anaerobic and aerobic digestion of complex organic materials.[5,6]

Lawrence and McCarty[7] assumed that the specific growth rate can be described by the Monod model, with little or no modification. They, for example, proposed an equation to predict the specific substrate utilization rate:

$$q = (q_m S)/(K_s + S) \tag{11.5}$$

where q_m is the maximum microbial specific substrate utilization rate. However, several others such as Grau *et al.*,[8] Grady and Williams[9] and Benfield and Randall,[10] found a few limitations in the Monod model, while predicting the volatile solids reduction and/or effluent concentration. They found that the Monod kinetic equation does not predict the dependence of effluent substrate concentration on influent substrate concentration. Grau *et al.*,[8] Grady and Williams,[9] McKinney[11] and Eckenfelder[12] all proposed different first-order kinetic equations in which the specific substrate utilization rate is proportional to the substrate concentration.

Although these first-order kinetic equations are simple and provide a good fit of experimental data, they fail to predict conditions of maximum biological activity, system washout, and product inhibition. Chen and Hashimoto,[13] as a result, adopted the Contois model to describe microbial specific growth rate. This model, unlike the Monod model, relates the bacterial specific growth rate to the limiting substrate concentration in the influent as well as in the effluent concentration:

$$\mu = [\mu_m S/S_0]/[K + (1 - K) S/S_0] \tag{11.6}$$

where μ_m is the maximum specific growth rate and K is a dimensionless kinetic parameter. This model can be used to describe methanogenesis from complex substrates expressed in terms of VS or COD concentrations over a wide range of retention times. Utilizing this model showed a good fit and a high level of predictability of experimental data.[13] Simplicity and versatility makes this model very useful for design applications.

For those substrates, such as volatile acids, that are inhibitory to the microorganisms at higher concentrations, the Monod kinetic equation was again found to be nonpredictable. Andrews[14,15] proposed an inhibition kinetic dynamic model, eqn (11.7), that can be used to predict digester instability due to product inhibition:

$$\mu = (\bar{\mu} S)/(K_s + S + S^2/K_i) \tag{11.7}$$

where K_i is the inhibition constant.

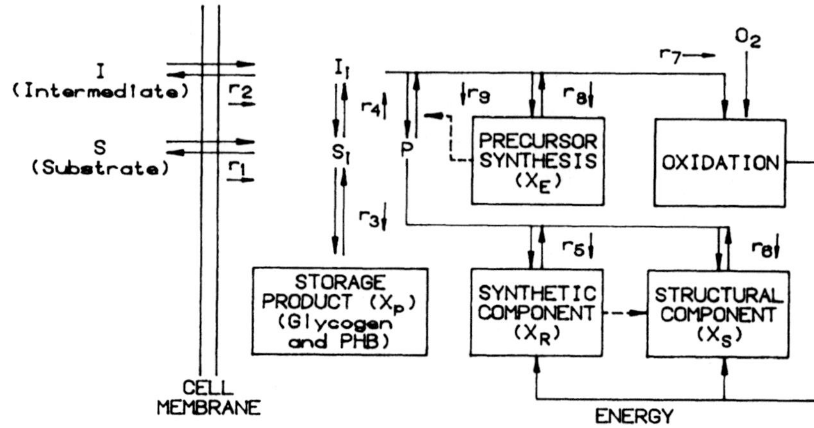

FIG. 11.1. Schematic representation of the comprehensive conceptual model for dynamic microbial growth.

Daigger and Grady[16] proposed a unifying dynamic theory on transient phenomena of microbial growth on soluble substrates based on the concepts of physiological adaptation and microbial species selection in the growth environment. In this theory, cellular mass was divided into five constituents: a synthetic component, a structural component, precursor molecules for those components, enzymes catalyzing the synthesis of the precursors, and storage and products. Figure 11.1 is the schematic representation of the conceptual model for dynamic microbial growth. According to Daigger and Grady[16] the transient phenomena such as growth responses and storage responses of the microbial growth can be modeled by proper selection of the rate constants.

11.3 COMPLETELY MIXED, CONTINUOUS-FLOW SYSTEM

Figure 11.2 is a schematic diagram of a completely mixed, continuous-flow digestion system. In that figure, S_{T0} is the total influent substrate concentration, S_T is the total effluent substrate concentration, Q is the influent or effluent flow rate, and X is the cell mass concentration in the effluent. Since the organic substrate is not completely digestible by the microorganisms, a portion of S_{T0} is refractory and will always be present in the effluent.[17] The biodegradable substrate concentrations are represented by S_0 and S, in the influent and effluent, respectively.

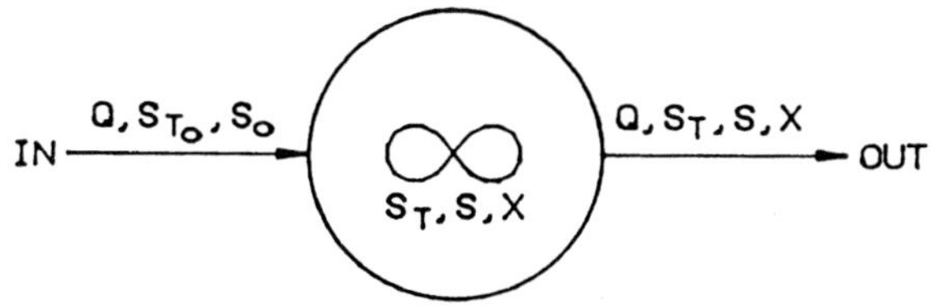

FIG. 11.2. Completely mixed, continuous flow system.

The mass balance of the microorganisms gives the following equation:

$$V(\mathrm{d}X)/(\mathrm{d}t) = -QX + \mu XV \qquad (11.8)$$

where X is the cell mass concentration in the digester, Q is the influent or the effluent slurry flow rate leaving the digester, V is the liquid volume of the digester, μ is the microbial specific growth rate, and t is time. Cell mass concentration of the predominant methane-producing bacteria in the influent is generally assumed to be negligible compared with the cell mass concentration in the digester.

Using the substrate mass balance of the system yields

$$V(\mathrm{d}s)/(\mathrm{d}t) = QS_0 - QS = 1/Y(\mu XV) \qquad (11.9)$$

For a continuously mixed and fed digester under steady-state, eqns (11.8) and (11.9) reduce to eqns (11.10) and (11.11), as follows:

$$\mu = 1/\theta \qquad (11.10)$$

$$\mu = Y(S_0 - S)/(\theta X) \qquad (11.11)$$

where θ is the hydraulic retention time that equals the ratio between V and Q. For a completely mixed system without solids recycling, cell mass concentration in the effluent should be the same as cell mass concentration in the digester. In this case, the microorganism retention time (θ_s), which is the ratio of the total cell mass in the digester to the cell mass in the effluent, is identical to hydraulic retention time.

From eqns (11.10) and (11.11), the cell mass concentration can be expressed as

$$X = Y(S_0 - S) \qquad (11.12)$$

In addition to understanding the anaerobic digestion system dynamically, the multistage approach requires the understanding of all the major steps involved in the fermentation process including the rate-limiting steps. Lawrence and McCarty,[7] using the Monod model, considered volatile fatty acids as growth-limiting substrate in studies on the kinetics of anaerobic digestion. (Their rationale was that the overall kinetics would be governed by the rate-limiting step in methanogenesis, which was the conversion of volatile fatty acids to methane and carbon dioxide.) Using eqns (11.2), (11.5), (11.10), and (11.11), they obtained the following two process kinetic equations for completely mixed, continuous-flow systems:

$$\theta^{-1} = Y_G q - K_d \tag{11.13}$$

and

$$S = [K_s(1 + K_d\theta)]/[\theta(Y_G q_m - K_d) - 1] \tag{11.14}$$

Chynoweth *et al.*[18] developed an expression of the Monod model that predicts steady-state methane yield and methane production rate as a function of the biodegradable organic loading rate of any select substrate concentration feedstock. This form is useful when the principal objective of the process is production of methane. The equations used to calculate the methane production rate (R_m) and methane yield (Y_m) can be presented as follows:

$$R_m = \alpha[S_0\bar{\mu}\theta - (S_0 + K_c)]/[\theta^2\bar{\mu} - \theta] \tag{11.15}$$

$$Y_m = \alpha[S_0\bar{\mu}\theta - (S_0 + K_c)]/[L\theta^2\bar{\mu} - L\theta] \tag{11.16}$$

where α is the methane yield coefficient, K_c is the saturation constant, and L is the organic loading rate. This model was tested for low- to medium-solids feedstocks (water hyacinth/sludge blend) in the laboratory. The model, with an example illustrated in Fig. 11.3, can be used to predict the steady-state digester performance at different organic loading rates, and to specify a range of loading rates for optimum digester performance.

Chen and Hashimoto,[17] using the Contois model, developed an equation to predict the steady-state effluent substrate concentration as follows:

$$S = KS_0/(\mu_m\theta - 1 + K) \tag{11.17}$$

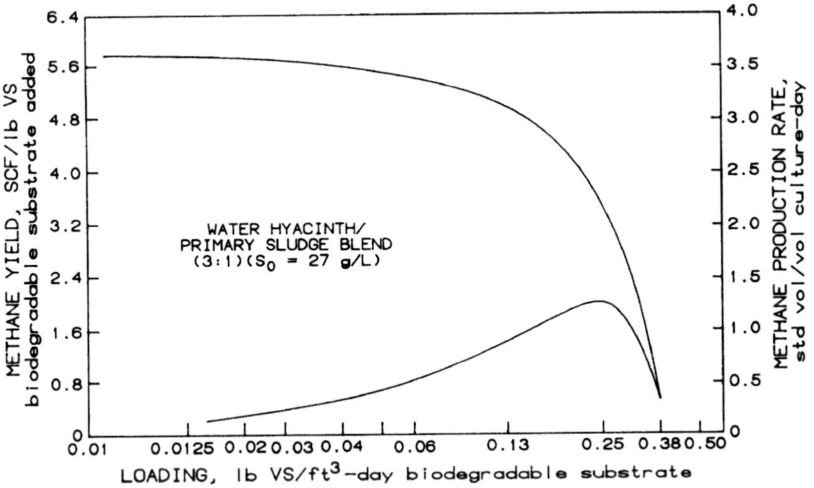

FIG. 11.3. Theoretical plots of methane yield and methane production rate versus biodegradable organic loading for 3:1 hyacinth/sludge blend.

This model shows that effluent substrate concentration is directly proportional to influent substrate concentration. It confirms the results of Grau *et al.*,[8] Grady and Williams,[9] and Benfield and Randall,[10] who state that the effluent quality obtained through biological treatment is a function of the influent substrate concentration.

Chen *et al.*[17,19] have further derived a kinetic equation for directly predicting methane yield and production rate under steady-state conditions:

$$Y_m = B_0\{1-[K/(\mu_m\theta-1+K)]\} \quad (11.18)$$

$$R_m = B_0L\{1-[K/(\mu_m\theta-1+K)]\} \quad (11.19)$$

where B_0 is the amount of methane production per unit of dry organic waste fed to the digester, Y_m is the ultimate methane yield of the waste, and R_m is the volumetric methane production rate. Equations (11.18) and (11.19) show that as hydraulic retention time increases, more organic waste is converted to methane. Production of methane increases as the maximum specific growth rate of the microorganism involved in producing methane increases and decreases as the kinetic parameter K increases. Also, the methane production and volumetric methane

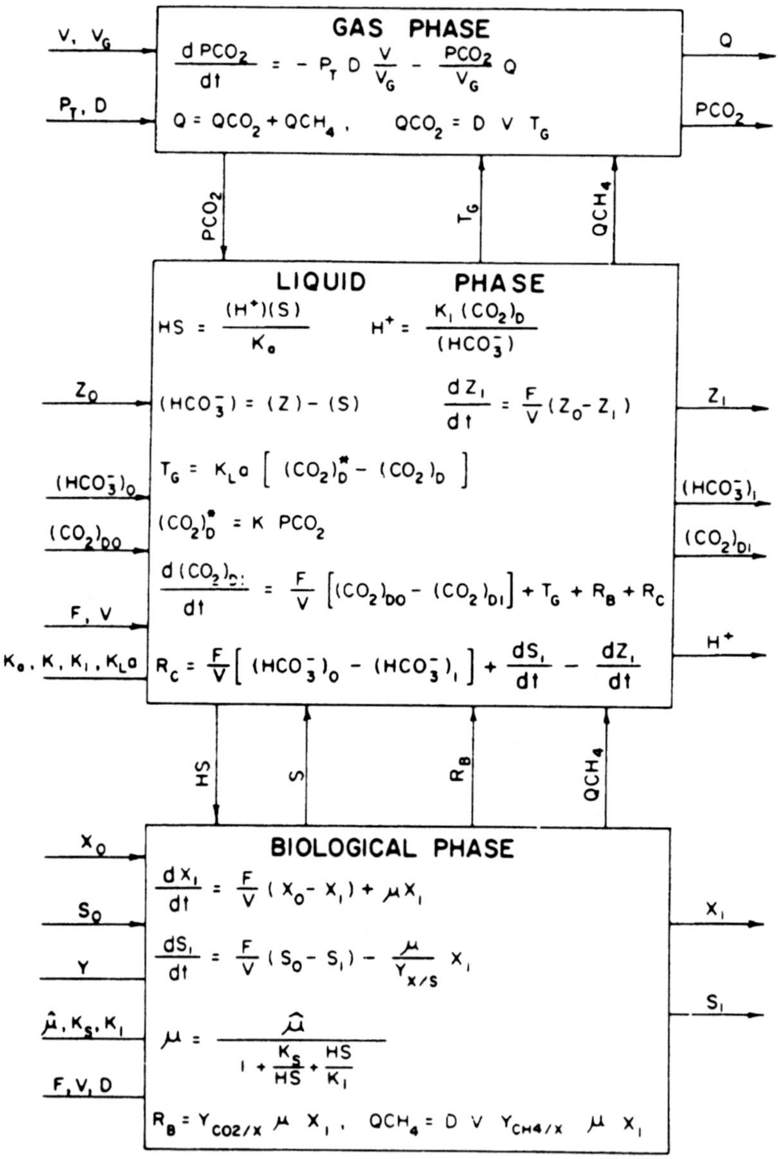

a

FIG. 11.4. Summary of mathematical model and information flow.[15]

$(CO_2)_D$	Dissolved carbon dioxide concentration (moles/liter)
$(CO_2)_D^*$	Concentration of dissolved CO_2 in the liquid phase when in equilibrium with the gas phase (moles/liter)
D	Conversion factor (liters gas/mole gas)
F	Liquid flow rate to reactor (liters/day)
H^+	Hydrogen ion concentration (moles/liter)
(HCO_3^-)	Bicarbonate concentration (moles/liter)
HS	Unionized substrate concentration (moles/liter)
K	Henry's law constant (moles/mm Hg-liter)
K_a	Ionization constant for acetic acid (moles/liter)
K_i	Inhibition constant (moles/liter)
K_La	Gas transfer coefficient (day^{-1})
K_s	Saturation constant (moles/liter)
K_1	Ionization constant for bicarbonate (moles/liter)
P_{CO_2}	Partial pressure of CO_2 in the gas phase (mm Hg)
P_T	Total pressure of CO_2 and CH_4 in the reactor (mm Hg)
Q	Total dry gas flow rate, $Q_{CH_4} + Q_{CO_2}$ (liters/day)
Q_{CH_4}	$DVY_{CH_4/X}\mu X_1$(liters/day)
Q_{CO_2}	$-D\,V\,T_G$ (liters/day)
R_B	Biological production rate of CO_2 (moles/liter-day)
R_C	Chemical production rate of CO_2 (moles/liter-day)
S	Substrate concentration (moles/liter)
t	Time (days)
T_G	Gas transfer rate (moles/liter-day)
V	Reactor liquid volume (liters)
V_G	Reactor gas volume (liters)
X	Organism concentration (moles/liter)
$Y_{CH_4/X}$	Yield constant (moles CH_4 produced/mole organisms produced)
$Y_{CO_2/X}$	Yield constant (moles CO_2 produced/mole organisms produced)
$Y_{X/S}$	Yield constant (moles organisms produced/mole substrate consumed)
Z	Net cation concentration (C − A, eq/liter)
μ	Specific growth rate (day^{-1})
$\hat{\mu}$	Maximum specific growth rate (day^{-1})
1	Subscript denoting reactor effluent
0	Subscript denoting reactor influent

b

FIG. 11.4—*contd.*

production rate are directly related to the ultimate methane yield. The maximum volumetric methane production rate ($R_{m_{max}}$) is given by

$$R_{m_{max}} = (B_0 S_{T0} \mu_m)/(1+\sqrt{K})^2 \qquad (11.20)$$

which occurs at

$$\theta = (1+\sqrt{K})/\mu_m \qquad (11.21)$$

These equations have been successfully used to describe methane production from anaerobic digestion of livestock wastes including beef cattle, dairy, and swine wastes and have subsequently been used for design calculations for methane production from organic waste by Lo *et al.*,[20] Chen,[21,22] Hill,[23] and Hashimoto *et al.*[24] The simplicity and accuracy of describing the methane production have made these equations a good choice when evaluating systems operation or studying systems inhibition by quantifying the parameter K.

Anaerobic digestion is a complex multistage dynamic process in which several physiologically different groups of bacteria work simultaneously.[25–28] For stable and effective digestion, these functional groups of bacteria should be in proper balance at every stage of anaerobic digestion. If this balance is disturbed for any reason or any inhibitory substance is present in the digester, the system can fail and may not rapidly recover. Andrews and Graef,[15] as mentioned earlier, developed a dynamic model to predict the state of this balance. This model, illustrated in Fig. 11.4, can be used to predict qualitatively the performance of anaerobic digesters during start-up, initial operation, and steady-state operation. Its key features are the use of an inhibition function (as shown in eqn 11.7) to relate volatile acid concentration and specific growth rate for the methane bacteria, consideration of the unionized acid as the growth-limiting substrate and inhibiting agent, and consideration of the interactions that occur in and between the liquid, gas, and biological phases of the digester. The model incorporating growth limitation and inhibition by the unionized volatile acid can be obtained by modifying the inhibition function (eqn 11.7) for pH values above 6, as follows:

$$\mu = \bar{\mu}\mathrm{HS}/[K_s + \mathrm{HS} + (\mathrm{HS})^2/K_i] \qquad (11.22)$$

where HS is unionized substrate concentration. Hill *et al.*[29] and Kleinstreuer and Poweigha[30] also described the system dynamics to predict qualitatively the systems failure; however, the rate constants are difficult to determine qualitatively.

11.4 COMPLETELY MIXED, CONTINUOUS-FLOW SYSTEM WITH SOLIDS RECYCLE

Figure 11.5 gives the schematic diagram for a completely mixed, continuous system with biological solids recycle. Lawrence and McCarty[7] assumed that all the substrate utilization occurred only in the digester because the separator volume was small and the recycle was continuous. The microbial retention time was given by

$$\theta_s = VX/[Q_R X_R + (Q - Q_R) X_e] \tag{11.23}$$

where Q_R is the flow rate of the fraction of the solids separator underflow that is recycled from the system, X_R is the biological solids concentration in the underflow from the solids separator, X_e is the solids concentration in the clarifier overflow from the solids separator, and Q is the flow rate of the incoming waste stream. Using a mass balance of biological solids and eqn (11.5), and replacing θ_s with θ, they obtained results similar to those observed with eqn (11.18) for the effluent substrate concentration. Gates *et al.*[31] developed an equation to predict effluent organism concentration (X_e) using a specific autodestruction rate (K_d) during steady state:

$$X_e = [Y_G(S_0 - S)]/[K_d + C] \tag{11.24}$$

where C is the recycle factor. Since there is no easy way to measure the concentration of cell mass when insoluble substrates are digested, the kinetic equations developed for solids recycling are not useful for the insoluble substrate digestion design.

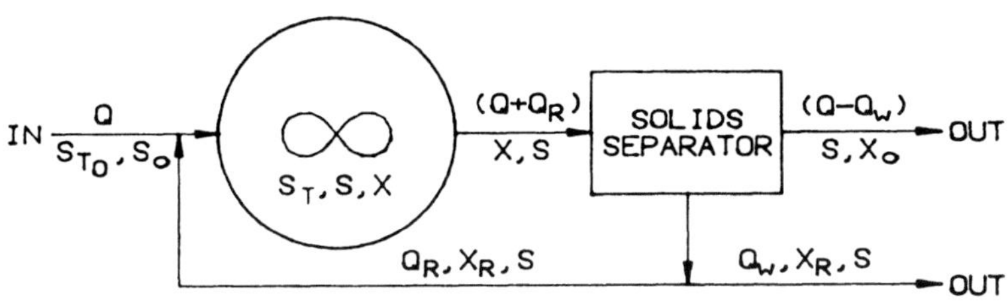

FIG. 11.5. Completely mixed, continuous system with solids recycle.

11.5 NONMIXED VERTICAL FLOW REACTOR

Nonmixed vertical flow reactor design does not employ mechanical mixing; however, it assumes some vertical mixing as a result of gas production. In this system, the substrate concentration decreases from the inlet to the outlet of the digester, and the digester can be operated either in upflow or in downflow mode depending upon the rheology of the substrate. This reactor design allows solids accumulation in the digester, which permits longer feed solids and microorganisms retention times than hydraulic retention time.

Srivastava and Chynoweth,[32] using the modified first-order kinetic model, developed kinetic equations that predict methane production rate and methane yield as a function of substrate concentration and hydraulic retention time. This model is very useful for practical application in the field, where the feed characteristics vary considerably from one season to the other. The equations used to calculate the methane production rate (R_m) and methane yield (Y_m) can be presented as follows:

$$R_m = \frac{\alpha(S_0 K_R \sigma) L}{(S_0 K_c \sigma + S_0)} \tag{11.25}$$

$$Y_m = \frac{\alpha(S_0 K_R \sigma)}{(S_0 K_c \sigma + S_0)} \tag{11.26}$$

where α is the methane yield coefficient, L is the organic loading rate, and K_R is the reaction rate constant. These equations have been successfully used to describe methane production and volatile solids destruction in a pilot-scale anaerobic digestion of different blends of biomass and wastes.[32,33]

11.6 TWO-PHASE FERMENTATION

Ghosh and Pohland,[34] using the Monod model, developed a process kinetic model for acidogenesis of soluble substrates in a completely mixed, continuous system under steady-state conditions. By controlling the hydraulic retention time so that there is washout of methanogenic bacteria, the physical separation of the two-phase fermentation is possible.

If the first phase is operated at a hydraulic retention time shorter than the retention time required by the methanogenic bacteria for their multiplication, then the liquid or acid production can be given by the following equation:

$$P_l = [a_l(Y_e^{-1} + M\theta)(S - S_0)]/[Y_e^{-1} + Y_p^{-1} + M\theta] \tag{11.27}$$

where a_l is the true liquid product yield constant of acidogenesis, Y_e is the energy yield coefficient, M is the maintenance coefficient for acid formers, Y_p is the biomass yield coefficient. The volume of gases produced by acidogenesis is given by

$$G = a_g V/\theta[(S_0 - S) - X_a Y_e^{-1}] \tag{11.28}$$

where V is the digester volume and a_g is the true gaseous product yield constant during acidogenesis.

11.7 PLUG FLOW SYSTEM WITH SOLIDS RECYCLE

Plug flow assumes that no longitudinal mixing occurs between adjacent fluid elements. In this system, the substrate concentration decreases along the length of the digester from the inlet to the outlet, while microbial mass concentration increases from its initial value as a result of the substrate uptake. The situation is illustrated in Fig. 11.6, which also shows the solids recycle from the solids separator.

For steady-state, the mass balance for a single substrate and culture system gives

$$V_x \mathrm{d}S/\mathrm{d}x = -qx \tag{11.29}$$

and

$$V_x \mathrm{d}X/\mathrm{d}x = Yqx \tag{11.30}$$

where x is the distance along the digester from the inlet and V_x is the flow speed of the substrate and the microorganism in the x direction.

To simplify the system so that the equation can be solved, Lawrence and McCarty[7] assumed that microbial concentration in the digester

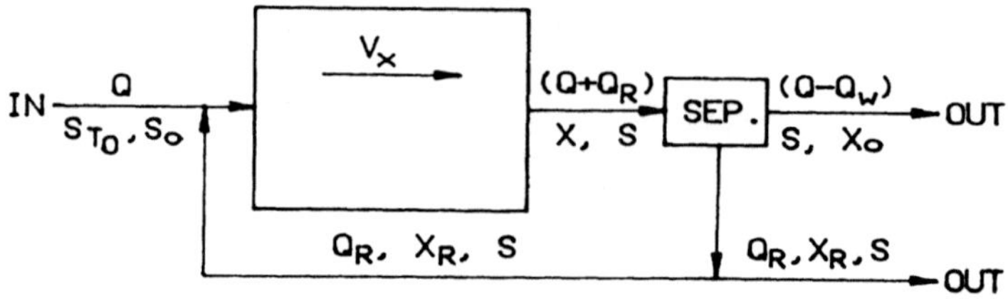

FIG. 11.6. Plug-flow system with solids recycle.

remained fairly constant. They have related the hydraulic retention time (θ) to the substrate concentration by

$$\theta = [1/(q_m X)][K_s \ln(S_i/S) + (S_i - S)] \quad (11.31)$$

where S_i is the total incoming biodegradable substrate concentration and S is the effluent biodegradable substrate concentration leaving the digester.

A relationship between effluent concentration and microbial solids retention time (θ_s) for a plug-flow system with solids recycle was also derived:

$$\frac{1}{\theta_s} = [Y_G q_m (S_0 - S)]/[(S_0 - S) + eK_s] - K_d \quad (11.32)$$

where

$$e = (1 + R)\ln[(RS + S_0)/(1 + K_d)] \quad (11.33)$$

and R is the ratio between Q_R and Q.

Since ideal plug flow does not exist and there is always backmixing and mixing by gas generation, the kinetic equations developed for completely mixed systems are sometimes used to describe anaerobic digestion using plug-flow systems, although this approach is not ideal and may lead to design and/or operational problems.

11.8 ANAEROBIC FILTER SYSTEM

In anaerobic filters, microorganisms exist in a biofilm attached to a medium. In an ideal anaerobic filter system it is assumed that the organic matter is removed from the stream by the biofilm.[35,36] It is also assumed that the organic matter, transported into the biofilm by molecular

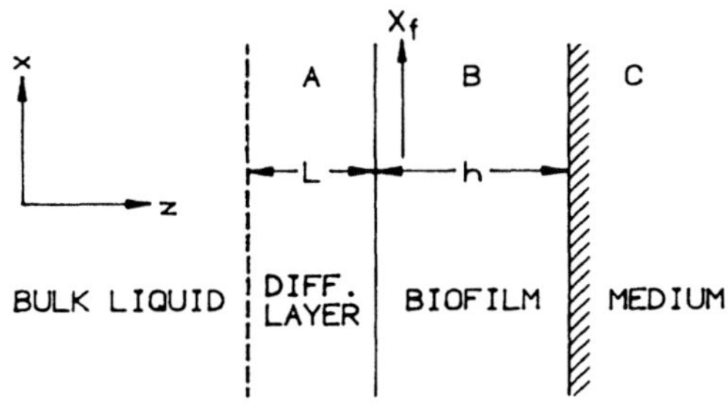

FIG. 11.7. Ideal biofilm.

diffusion, is subsequently converted by the biofilm-associated organisms. Figure 11.7 illustrates the ideal biofilm, which has a uniform cell density of X_f and uniform thickness of h. Substrate concentrations change only along the z direction.

It is generally assumed that the substrate utilization at any point within the biofilm follows the Monod kinetics, and the substrate transport in the biofilm is generally described by molecular diffusion.[35,36] Thus, under steady-state, at any point in the biofilm, the substrate transport in the biofilm should be equal to the substrate utilization, so

$$D_f(d^2S_f)/(dZ^2) = (q_m X_f S_f)/(K_s + S_f) \qquad (11.34)$$

where X_f is the cell mass concentration in the biofilm, S_f is the substrate concentration in the biofilm, and D_f is the substrate diffusivity in the biofilm. Assuming that $S_f >> K_s$, X_f is the constant in the biofilm, and the substrate flux through the biofilm surface is calculated by Fick's first law of molecular diffusion, DeWalle and Chian[35] obtained an equation for substrate flux (J) as a function of bulk substrate concentration (S_b):

$$J = (2q_m D_f X_f S_b)^{1/2} \qquad (11.35)$$

An equation was obtained for substrate utilization rate per unit volume ($\dot{F}$) as follows:

$$\dot{F} = k_1(A/V)(S_b)^{1/2} \qquad (11.36)$$

where k_1 is the coefficient based on zero-order kinetics and A/V is the ratio of the surface area of the biofilm to the digester volume. They concluded that if $S_f << K_s$, the substrate utilization rate ($\dot{F}$) is directly proportional to the bulk substrate concentration.

In order to apply growth kinetics to a steady-state biofilm, Rittman and

McCarty[36] assumed that the rate of energy captured through substrate utilization and cell growth is equal to the required rate of energy expenditure for cell maintenance. The thickness of a biofilm at steady-state is therefore determined by the flux of the substrate into the biofilm and the bacterial growth and decay rate. It is assumed that the total amount of biofilm mass is just equal to that which can be supported by the substrate flux, so

$$h = JY_G/(K_d X_f) \tag{11.37}$$

where h is the thickness of a biofilm under steady state.

Dahab and Young[37] reported that one-half to two-thirds of the biological solids mass in the lower one-third of the biofilm fixed-bed reactor was not attached to the media matrix. Analyses of the activity of these solids showed that these cell masses were very active. This suggests that the microbial activities are not limited to the cell mass attached to the media. Trulear and Characklis[38] considered the net growth of biofilm by considering continuous microbial detachment during the digestion process. They indicated that the amount of detached microorganisms depends upon the hydrodynamic conditions and sloughing caused by nutrient depletion deep within the biofilm. From this theory, they developed a logistics equation that can be used to determine substrate removal, biofilm thickness, and suspended biological mass in the system:

$$Z = Z_0 e^{kt}/[1 - (Z_0/Z_m)(1 - e^{-kt})] \tag{11.38}$$

where Z can be substrate removal, biofilm thickness or suspended biomass concentration in the reactor, Z_0 is Z at $t = 0$, Z_m is the maximum value attained by Z and k is the rate constant.

Grady *et al.*,[39] in their state-of-the-art review of the fixed-film systems, identified a few limitations of existing models and proposed a few research areas that need attention before a complete understanding of the system can be achieved. These research areas include the determination of the type of biofilm; the liquid interface that exists and its effect on liquid transportation in the biofilm; modeling for biofilm thickness; biofilm loss by sloughing; determination of variation in density of biofilm and its effect on substrate uptake rate; further evaluation of determinants of the concentration profiles within the biofilm; and more work on intrinsic kinetics under double substrate limitations.

11.9 FLUIDIZED-BED SYSTEM

In fluidized-bed systems, as with anaerobic filters, microorganisms exist in a biofilm attached to a nondegradable medium. The organic matter is transported into the biofilm where it is utilized. A kinetic model for a plug-flow, fluidized-bed reactor was developed by Shieh.[40] He assumed that the kinetics of the substrate uptake is of zero order and that there is a reaction-free zone on the surface of the medium. Based on these assumptions, he obtained the following equation for the substrate uptake rate (R) in the biofilm that is related to the reaction rate occurring within the biofilm:

$$R = \eta_0 \rho k(1 - E)(1 - r_m/r_p)^3 \tag{11.39}$$

where η_0 is the effectiveness factor, E is the porosity, ρ is the biofilm dry density, k is the zero-order rate constant, r_m is the radial distance measured from the center of the media, and r_p is the bioparticle radius. The effectiveness factor, η_0, can be calculated as

$$\eta_0 = [1 - (r/r_p)^3]/[1 - (r_m/r_p)^3] = \{[3{\cdot}012(\rho k/D_f)^{-0{\cdot}45}(r_p)^{-0{\cdot}9}] /[1 - (r_m/r_p)^3]\}\, S_b^{0{\cdot}45} \tag{11.40}$$

The following equation was developed to predict substrate concentration as a function of residence time and feed substrate concentrations:

$$S_b^{0{\cdot}55} = -K(Z/V_Z) + S_0^{0{\cdot}55} \tag{11.41}$$

where

$$K = 1{\cdot}657(1 - E)(\rho k)^{0{\cdot}55}[(D_f^{0{\cdot}45})/(r_p^{0{\cdot}9})] \tag{11.42}$$

and Z/V_z is the residence time.

Rittman[41] developed a unified model of biofilm reactor kinetics based on substrate-utilization kinetics, mass transport, biofilm growth, and reactor analysis. This model is based on common principles, which made the model useful for complete-mix, fixed-bed, and fluidized-bed reactors with and without recycle. Rittman[41] presented an equation for reactor modeling that incorporates the substrate utilization and biofilm growth:

$$E(\partial S)/(\partial t) = -v(\partial S)/(\partial x) + D_H(\partial^2 S)/(\partial x^2) - aJ - (\epsilon k X_f S)/(K_s + S) \tag{11.43}$$

where v is the superficial flow velocity, a is the surface area per reactor volume, and D_H is hydrodynamic diffusivity. He also presented an equation to determine average flux ($\bar{J}$) specially in fluidized-bed reactors:

$$\bar{J} = (1/V) \int_0^V J\,dV = [(S_0 - S)\theta]/[aV] \qquad (11.44)$$

where dV is the differential volume of the reactor. Unlike the models based on loading criteria, this unified model can predict the performance of a variety of biofilm processes because it incorporates the interactions among substrate utilization, biofilm growth, and the reactor configuration. Nevertheless, some important limitations of the modeling results need to be considered before directly applying this model. These limitations include non-steady-state operation, dual-substrate limitation, bed expansion, wall effects, clogging, and shearing.

11.10 SUMMARY

Process kinetics and modeling are important tools for reactor design, process scale-up, specification of operating conditions, prediction of anaerobic digester performance, stability, and effluent quality at specified operating conditions. Ideally, process models should be able to describe the qualitative and quantitative aspects of microbial reactions ranging from mass transfer to population dynamics in different reactor configurations under different operational and environmental conditions. However, the task of obtaining valid required kinetic constants is complicated by the fact that anaerobic digestion is a complex multistage dynamic process involving the concerted effort of several groups of bacteria, the composition of which varies in an unknown manner with changes in retention time,[25–28] feedstock, temperature, reactor type, and other operating conditions. It is further complicated by the lack of a valid and reliable method for quantification of the microbial cell mass in digesters containing insoluble substrates. The detail to which modeling of anaerobic digestion is possible is limited by the lack of knowledge on the specific bacteria involved, their metabolism and physiological limitations, and the interactions between populations of the organisms. Current modeling efforts are, therefore, limited to inputs and outputs representative of broad categories of microbial activity representing the concerted activities of many unidentified species of bacteria. As we learn more about the specific organisms involved and their interactions it should be possible to increase the level of sophistication of the models and their capacity to predict performance accurately.

In general, most of the existing models for anaerobic digestion apply to steady-state conditions in CSTR reactors and employ Monod kinetics to describe the relationship between bacterial growth, substrate utilization, and product formation. Research is needed on models for new digester types, non-steady-state conditions, and also on those models that describe the relationship between performance and a wide variety of operation and performance parameters. To a large extent, development of such models is dependent upon development of knowledge on the fundamental reactions in the system.

A variety of different reactor configurations are being developed for improved performance on biomass and waste feedstocks. In particular, batch, nonmixed continuous feed, attached-film, and multiphase digesters will be challenges for future modeling efforts. Dynamic models are needed that describe the many non-steady-state conditions that prevail in anaerobic digestion, such as start-up, change in loading and retention time, inhibition, and recovery from inhibitory conditions. Modeling efforts should be expanded to include more variables such as pH, alkalinity, oxidation–reduction potential, volatile acids, gas composition, and gas production. Such comprehensive models could lead to automated process-evaluation and control.

There is a need to model limiting reactions within the digestion process. As more information is gained on the hydrolysis reactions, efforts should be made to model factors limiting hydrolysis such as feed composition, particle size, etc. Microbial populations within the methane fermentation are differentially limited by macro- and micronutrients. As these relationships become identified they should be incorporated into comprehensive process models.

The anaerobic digestion process is largely designed and operated on the basis of empirically derived information. Consequently, digesters are frequently overdesigned and subject to unpredictable failure in performance. The use of comprehensive dynamic models that continue to incorporate fundamental knowledge on factors limiting the performance of the process should lead to more economic design and predictable stable performance.

REFERENCES

1. Lawrence, A. W. and McCarty, P. L., 'Kinetics of Methane Fermentation in Anaerobic Treatment,' *J. Wat. Poll. Contr. Fed.* **41**:2, 1969.
2. Kirsch, E. J. and Sykes, R. M., 'Anaerobic Digestion in Biological Waste Treatment,' in D. J. D. Hockenhull, Ed., *Progress in Industrial Microbiology*. London: J. and A. Churchill, 1971.
3. Pirt, S. J., 'The Maintenance Energy of Bacteria in Growing Cultures,' *Proc. Roy. Soc. London, Ser. B* **163**: 224, 1965.
4. Monod, J., 'The Technique of Continuous Culture Theory and Applications,' *Annales de l'Institut Pasteur* **79**:390, 1950.
5. Andrews, J. F., Cole, R. D. and Pearson, E. A., 'Kinetics and Characteristics of Volatile Acid Production in Anaerobic Fermentation Processes,' *Int. J. Air Water Poll.* **9**:439, 1965.
6. O'Rourke, J. R., 'Kinetics of Anaerobic Treatment at Reduced Temperatures,' PhD Thesis, Stanford University, 1968.
7. Lawrence, A. W. and McCarty, P. L., 'Unified Basis for Biological Treatment Design and Operation,' *J. of San. Eng. Div.* **SA3**:757, 1970.
8. Grau, P., Dohanyos, M. and Chudoba, J., 'Kinetics of Multicomponent Substrate Concentrate Removal by Activated Sludge,' *Wat. Res.* **9**:637, 1975.
9. Grady, C. P. L., Jr and Williams, D. R., 'Effects of Influent Substrate Concentration on the Kinetics of Natural Microbial Populations in Continuous Cultures,' *Wat. Res.* **9**:171, 1975.
10. Benfield, D. and Randall, C. W., 'Evaluations of a Comprehensive Kinetic Model for the Activated Sludge Process,' *J. Wat. Poll. Contr. Fed.* **49**:7, 1636, 1977.
11. McKinney, R. E., 'Mathematics of Complete Mixing Activated Sludge,' *J. San. Eng. Div., Amer. Soc. Civ. Engrs*, **88**:87, 1962.
12. Eckenfelder, W. W., Jr, 'Mathematical Formulation of the Biological Oxidation Process,' In W. W. Eckenfelder and J. McCabe, Eds, *Advances in Biological Waste Treatment*. New York: Pergamon Press, 1963.
13. Chen, Y. R. and Hashimoto, A. G., 'Kinetics of Methane Fermentation,' *Biotech. and Bioeng.* **8**:269, 1978.
14. Andrews, J. F., 'A Mathematical Model for the Continuous Cultivation of Microorganisms Utilizing Inhibitory Substances,' *Biotech. and Bioeng.* **10**:707, 1968.
15. Andrews, J. F. and Graef, S. P., 'Dynamic Modeling and Simulation of the Anaerobic Digestion Process,' In *Anaerobic Biological Treatment Processes, Adv. in. Chem. Ser.* **105**:126, Washington, DC, 1971.
16. Daigger, G. T. and Grady, C. P. L., Jr, 'The Dynamics of Microbial Growth on Soluble Substrates: A Unifying Theory,' *Wat. Res.* **16**:365, 1982.
17. Chen, Y. R. and Hashimoto, A. G., 'Substrate Utilization Kinetic Model for Biological Treatment Processes,' *Biotech. Bioeng.* **22**:2081, 1980.
18. Chynoweth, D. P., *et al.*, 'Kinetics and Advanced Digester Design for Anaerobic Digestion of Water Hyacinth and Primary Sludge,' Paper presented at the *Fourth Symposium on Biotechnology in Energy Production and Conservation*, Gatlinburg, Tenn., 11–14 May 1982.

19. Chen, Y. R., Varel, V. H. and Hashimoto, A. G., 'Methane Production From Agricultural Residues,' *J. of Indust. Eng. and Chem.* **19**, 12:471, 1980.
20. Lo, K. V., Carson, W. M. and Jeffers, K., 'A Computer-Aided Design Program for Biogas Production From Animal Manure,' p. 133 in *Livestock Wastes: A Renewable Resource*. St Joseph, Mich.: ASAE, 1981.
21. Chen, Y. R., 'Kinetic Analysis of Anaerobic Digestion of Pig Manure and Its Design Implications,' *Agric. Wastes* **8**:65, 1983.
22. Chen, Y. R., 'Biogas Digester Design,' Chapter 2 in D. L. Wise, Ed., *Fuel Gas Systems*. Boca Raton, Fla: CRC Uniscience Publications, 1983.
23. Hill, D. T., 'Design of Digestion Systems for Maximum Methane Production,' *Trans. of the ASAE* **25**,1:226, 1982.
24. Hashimoto, A. G., Chen, Y. R. and Varel, V. H., 'Theoretical Aspects of Anaerobic Fermentation: State-Of-The-Art,' p. 86 in *Livestock Wastes: A Renewable Resource*. St Joseph, Mich.: ASAE, 1981.
25. Bryant, M. P., 'The Microbiology of Anaerobic Degradation and Methanogenesis With Special Reference to Sewage,' in H. G. Schlegel, Ed., *Symposium on Microbial Energy Conversion*. Göttingen, Federal Republic of Germany: E. Goltze KG, 1976.
26. McCarty, P. L., 'Anaerobic Waste Treatment Fundamentals,' *Pub. Works* **95**:91, 123, 1964.
27. Wolfe, R. S., 'Microbial Biochemistry of Methane — A Study in Contrasts,' in J. R. Quayle, Ed., *Microbial Biochemistry*, Vol. **21**. Baltimore: University Park Press, 293, 1979.
28. Zeikus, J. G., 'Microbial Populations in Anaerobic Digesters,' Paper presented at *First International Symposium on Anaerobic Digestion*, University Industry Centre, University College, Cardiff, Wales, 17–21 September 1983.
29. Hill, D. T., Tollner, E. W. and Holmberg, R. D., 'The Kinetics of Inhibition in Methane Fermentation of Swine Manure,' *Agric. Wastes* **5**:2, 105, 1983.
30. Kleinstreuer, C. and Poweigha, T., 'Dynamic Simulator for Anaerobic Digestion Processes,' *Biotech. Bioeng.* **24**:293, 1979.
31. Gates, W. E., Smith, J. H., Lin, S. D. and Ris, C. H., III, 'A Rational Model for the Anaerobic Contact Process,' *J. Wat. Poll. Contr. Fed.* **39**:1951–1970, December 1967.
32. Srivastava, V. J. and Chynoweth, D. P., 'Kinetic Analyses of Biogasification of Biomass/Waste Blends and its Engineering Significance,' in *Proceedings: Energy From Biomass and Wastes X*, Institute of Gas Technology, Washington, DC, April 1986.
33. *Hyacinth Wastewater Treatment Workbook (Draft)*, prepared by WED Enterprises, Glendale, Calif., and Black and Veatch, Engineers–Architects, Kansas City, Mo., for Gas Research Institute, Chicago, January 1986.
34. Ghosh, S. and Pohland, F. G., 'Kinetics of Substrate Assimilation and Product Formation in Anaerobic Digestion,' *J. Wat. Poll. Contr. Fed.* **46**, 4, 748, 1974.
35. DeWalle, F. B. and Chian, E. S., 'Kinetics of Substrate Removal in a Completely Mixed Anaerobic Filter,' *Biotech. and Bioeng.* **18**:1275, 1976.
36. Rittman, B. E. and McCarty, P. L., 'Model of Steady-State-Biofilm Kinetics,' *Biotech. and Bioeng.* **22**:2359, 1980.

37. Dahab, M. F. and Young, J. C., 'Retention and Distribution of Biological Solids in Fixed-Bed Anaerobic Filters,' Paper published in *Proc. First-International Conference on Fixed-Film Biological Processes*, Kings Island, Ohio, April 1982.
38. Trulear, M. G. and Characklis, W. G., 'Dynamics of Biofilm Processes,' *J. Wat. Poll. Contr. Fed.* **54**:1288–1301, 1982.
39. Grady, C. P. L., Jr, Harlow, L. F. and Riesing, R. R., 'Effects of Growth Rate and Influent Substrate Concentration on Effluent Quality From Chemostats Containing Bacteria in Pure and Mixed Culture,' *Biotech. and Bioeng.* **14**:391, 1972.
40. Shieh, W. K., 'Suggested Kinetic Model for the Fluidized-Bed Biofilm Reactor,' *Biotech. and Bioeng.* **22**:2359, 1980.
41. Rittman, B. E., 'Comparative Performance of Biofilm,' *Biotech. and Bioeng.* **24**:1341–70, 1982.

12
Net Energy

VIPUL J. SRIVASTAVA
Institute of Gas Technology, Chicago, Illinois, USA

12.1 INTRODUCTION

Until recently little attention was given to net energy in anaerobic digestion because this process was used primarily for treatment of municipal and industrial wastes. Generalizations were for energy requirements with little regard to the differences in digester design and operational parameters.[1–3] However, with recent emphasis on the use of this technology for conversion of organic matter to energy products, there is now a greater need to define net energy for each type of digestion system and to evaluate the factors that affect the net energy and energy efficiencies. Depending upon the feed characteristics and operational parameters, the total energy input in anaerobic digestion can vary from less than 10% to more than 100% of the product energy. Several recent publications focused on this subject and critically reviewed some variables, such as digester temperature, feed solid content, and loading, that affect the energy requirement during anaerobic digestion.[4–6] Major categories of energy requirements include feed heating energy, digester heat loss, pumping energy, mixing energy, and miscellaneous utilities.

12.2 FEED HEATING ENERGY

This category includes energy required to raise the temperature of raw feed to the digester operating temperature. Three factors influence the feed heating energy requirement: ambient temperature, digester operating temperature, and feed total solids content.

12.2.1 Ambient Temperature

Generally, the temperature of incoming raw feed is governed by the ambient temperature of that location. In the United States it can change from a year-round average of 2.5 to 24.5 °C.[7] Such a wide variation in ambient temperature can significantly change the energy requirement for identical plants from one location to another. This suggests that when determining the energy requirement for any specific digester, local ambient temperature rather than assumed temperature should be considered.

12.2.2 Digester Operating Temperature

Anaerobic digesters are generally operated either at mesophilic (35 °C) or thermophilic (55 °C) temperature, but digestion temperatures can range from ambient temperature to about 70 °C, depending on the type of feedstock and the location of the plant. Table 12.1 shows estimated values of major types of energy requirements for different examples of digestion systems. For a feed slurry of 15 wt% total solids (TS) content, the feed heat requirement for mesophilic digestion (Example IV) is only 7–8% of the product energy, whereas for thermophilic digestion (Example V) it is 13–15% of the product energy. This suggests that digester operating temperature can also affect energy requirements significantly.

12.2.3 Feed Total Solids (TS) Content

The TS of the raw feed has the most significant effect on the feed heat requirement. This is mainly because the specific heat of the solids is low compared with that of water. The feed slurry, which is anaerobically digested, can vary in TS content from less than 1 wt% to more than 50 wt% (with water recycle within the digester system). Such a wide variation can significantly change feed heat requirements. As shown in Table 12.1, the feed heat requirement for mesophilic digestion is 100–110% of the total energy produced when 1 wt% of TS feed slurry is used, while it is only 7–8% when 15 wt% of the TS feed slurry is used. This explains the reason why the ambient temperature digester is often preferred for very dilute feedstocks. Figure 12.1 also illustrates the effect of feed solids content on feed heat requirements for mesophilic digestion and the percent of product energy required at different values of methane yields.[4]

The equation to calculate the feed heat requirement (Q_i), as given by

TABLE 12.1

Energy Requirement Estimates for Various Components of Conventional Digesters* (% of Energy Output) [60 wt% VS-Content Feed, 15 °C Feed Slurry Temperature, 5 SCF/lb VS added (0·31 SCM/kg VS added)-Methane Yield]

Example No.	*Digestion system*	*Feed heat requirement*	*Digester heat loss*	*Mixing & pumping*	*Miscellaneous utilities*	*Total energy input*
I	Mesophilic digestion, 1 wt% TS content feed 0·1 lb VS/ft^3-day (1·6 kg VS/m^3-day) loading	100–110	7–8	2–3	2–3	110–125
II	Mesophilic digestion, 5 wt% TS content feed 0·1 lb VS/ft^3-day (1·6 kg VS/m^3-day) loading	22–25	7–8	2–3	2–3	35–40
III	Mesophilic digestion, 5 wt% TS content feed 0·5 lb VS/ft^3-day (8·0 kg VS/m^3-day) loading	22–25	2·5–3·5	1–1·5	0·8–1	26–31
IV	Mesophilic digestion, 15 wt% TS content feed 0·5 lb VS/ft^3-day (8·0 kg VS/m^3-day) loading	7–8	2·5–3·5	1–1·5	0·8–1	11–14
V	Thermophilic digestion, 15 wt% TS content feed 0·5 lb VS/ft^3-day (8·0 kg VS/m^3-day) loading	13–15	8–10	1–1·5	0·8–1	23–28
VI	Thermophilic digestion, 15 wt% TS content feed 1·0 lb VS/ft^3-day (16 kg VS/m^3-day) loading	13–15	4–5	1–1·5	0·6–0·8	19–22

*This table was prepared from the information presented in Ref. 5.

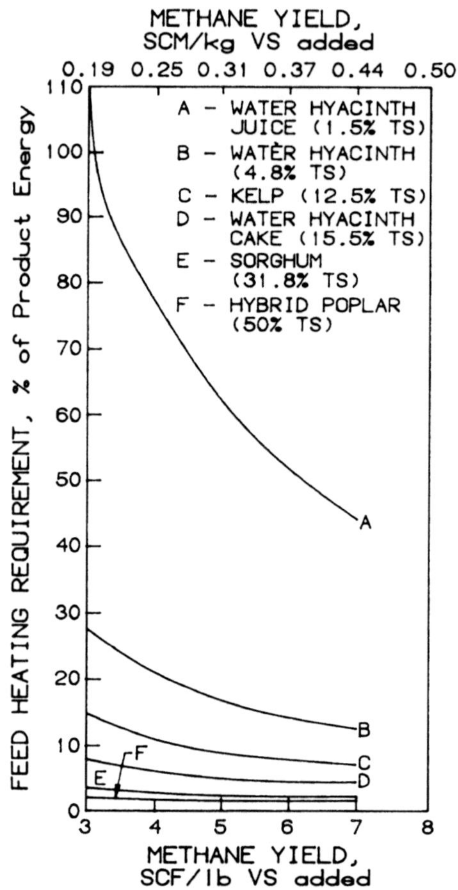

FIG. 12.1. Feed heating requirements for mesophilic digestion of various feedstocks (raw feed temperature 15 °C).

Chen and Hashimoto,[6] is

$$Q_i = WC_p(T_D - T_A) \tag{12.1}$$

where W = total weight of the feed slurry added per day, C_p = the specific heat of the feed slurry, T_D = the digester operating temperature, and T_A = the ambient temperature. The specific heat (C_p) of each feedstock is different and should be calculated individually for each biomass source.

The feed heat requirement is usually the largest fraction of total energy requirement for anaerobic digestion. However, this can be reduced by heat exchange between the effluent and influent.

12.3 DIGESTER HEAT LOSS

Digester heat loss is the equivalent energy that is required to compensate for the loss of heat from the digester to its surroundings. There are four factors that influence heat loss: ambient temperature, digester operating temperature, construction material of the digester, and loading rate.

12.3.1 Ambient Temperature

The ambient temperature is generally lower than the digester operating temperature, which causes a driving force to develop for heat transfer (loss) from the digester to its surroundings. The lower the ambient temperature the more heat that will be lost, which suggests, as discussed above, that the actual local temperature should be considered when calculating the heat loss from the digester.

12.3.2 Digester Operating Temperature

As mentioned above, the driving force is developed due to the difference in digester and ambient temperatures. The higher the digester temperature, the more heat will be lost from the digester. As shown in Table 12.1, the heat loss from mesophilic digesters (Example IV) is 2.5–3.5% of the product energy, whereas from a thermophilic digester (Example V) it is 8–10% of the product energy.

12.3.3 Material of Construction

The heat transfer rate (loss) is different for different types of materials.[8] Thus, depending upon the type of material used for construction, the heat loss can be slightly different. Proper insulation and underground digester placement can reduce heat loss from the digester.

12.3.4 Loading Rate

The loading rate determines the culture volume required to process certain amounts of feedstock. For example, as the loading is increased, the digester size required for the operation will be decreased, which will result in lower surface area for heat loss and lower heat loss from the digester. This can be explained quantitatively from the examples given in Table 12.1. The heat loss from a thermophilic digester operated at 16 kg

VS/m^3-day loading (Example VI) is only 4–5%, whereas it is 8–10% when the digester is operated at 8.0 kg VS/m^3-day (Example V). Similar conclusions can be found from mesophilic digesters (Examples II and III) operating at two different loadings. It should be noted, however, that while the heat loss per unit surface area will decrease as the loading is increased, the heat loss per unit digester volume will increase due to a greater surface-area-to-volume ratio. This suggests that the heat loss from a larger digester will be lower than the heat loss from several small digesters having the same total volume.

The digester heat loss (Q_L) can be calculated by

$$Q_L = UA(T_D - T_A) \tag{12.2}$$

where U = overall heat transfer coefficient, A = surface area, T_D = digester operating temperature (or digester wall temperature), and T_A = ambient temperature.

The digester heat loss is usually the second largest fraction of the total energy requirement for the operation of an anaerobic digester. It can be minimized by using proper insulation and also locating the digesters next to each other if more than one digester is required for the total operation.

12.4 PUMPING ENERGY

Throughout the digestion scheme various types of materials have to be transported from one place to another. Several of these tasks such as digester feeding, effluent withdrawal, solids recycle, gas recirculation, etc., are carried out by pumps that require energy for operation. Pumping energy usually depends on the solids content and the rheological properties of the slurry.

12.4.1 Solids Content

The pumping energy is influenced by the solids content of the slurry because slurry with high solids requires larger clearance in the pump, which increases the slip and reduces the mechanical efficiency. Also, due to high solids content, more power is required to overcome greater friction loss.

12.4.2 Rheological Properties

The rheological properties such as flow characteristics, viscosity, solids settling characteristics, etc., of the slurry have a significant effect on the pumping energy. A slurry with low viscosity and low solids settling characteristics will have low pumping energy requirements. On the other hand, a slurry with high viscosity and high solids settling characteristics will require substantially higher pumping energy due to increased friction losses and maintenance of high liquid velocity to prevent settling in the pipe.

Biomass differs substantially in solids and rheological properties, which results in different pumping characteristics and different energy requirements for different types of feedstocks. For example, several pumping tests conducted by IGT[9,10] for various purposes showed that kelp with a 12 wt% TS can be pumped very easily, whereas water hyacinth with a 12 wt% TS will require special techniques and more energy for transport. This difference is due mainly to the flow characteristics of those materials. The sea kelp is very slimy and has a low viscosity, while water hyacinth is very abrasive and high in fibers, which causes greater friction losses.

The pumping energy in a conventional digester is generally a small portion (less than 2%) of the product energy.[5] However, this requirement will change significantly in some of the unconventional digester designs. For example, the pumping energy in fluidized or expanded beds and upflow-sludge blanket digesters can be significantly higher than in other digesters because a continuous recirculation of culture is required during the normal operation of these digesters.

The objective of culture recirculation in fluidized or expanded bed digesters is to fluidize or expand the bed of support medium to facilitate bacterial contact with the substrate. If the specific gravity of the medium is high (such as 2.65, for sand), the friction pressure loss through the medium will result in a head loss of 1 m/m of bed depth or greater.[11] In order to overcome such a loss, additional energy is required for pumping. This additional energy also depends upon the degree of fluidization required, the cross-sectional area of the bed, and the physical characteristics of the slurry. A detailed analysis of total energy requirement in such a digestion scheme with regard to the difference in feed characteristics and support medium should be conducted before a valid assessment of net energy can be made.

12.5 MIXING ENERGY

High-rate conventional digesters and some unconventional digesters require mixing to maintain homogeneity in the tank and to reduce, if not eliminate, the scum formation caused by low specific gravity substances. Mixing is normally carried out by either mechanical agitation, digester contents recirculation, or gas recirculation. Thus, the mixing energy is influenced mainly by the fluid property (physical and chemical characteristics) of the digester content and the type of mixing equipment or technique employed.

12.5.1 Fluid Property

As mentioned in Section 12.4, biomass differs substantially in rheology and other physical properties, and this causes the mixing energy to be different in each case. For example, a slurry with high solids content and high settling characteristics will require more mixing to keep the particles in suspension. This will increase the energy requirement. Similarly, if the fluid is viscous, the power required to overcome frictional losses in any type of mixing equipment and the resistance in gas recirculation through the digester liquor will be greater, resulting in a higher mixing energy requirement.

12.5.2 Mixing Equipment

The mixing energy requirement is mainly influenced by the type of mixing equipment used. The selection of mixing equipment for any digester depends on several factors, including the size of the digester and the degree of mixing required. As discussed in Chapter 10, there are three major types of mixing equipment that are commonly used in this field: mechanical mixing, hydraulic mixing, and the mixing caused by digester gas recirculation. The methods to calculate energy requirement for each type of system are different. For mechanical mixing, several simple equations have already been worked out for various types of mixing-blade geometry. However, in most cases, these equations assume liquids having Newtonian fluid properties, and for digester contents that are often known to be non-Newtonian, the apparent viscosity has first to be determined before this method of determining energy requirements can be applied.

For hydraulic mixing (digester content recirculation) the energy required varies widely according to systems operating principles and flow rates. A minimum flow rate is usually about one complete volume turnover per day.

The power requirement for digester gas recirculation to achieve desirable agitation depends upon the amount of gas necessary to be discharged and, unlike hydraulic mixing where the amount of liquid recirculation can be estimated, the amount of gas to be diffused for efficient mixing is difficult to predict. It depends on many factors, such as tank geometry, type and position of diffusers, and the frequency of diffusion.

The mixing energy in conventional digesters is generally a small portion (less than 2%) of the product energy. As mentioned earlier, this amount will be different for different systems. In some unconventional digestion schemes, the mixing energy can be even less. For example, in an upflow solids reactor or plug-flow reactor, no external mixing is required for normal operation. This suggests that when calculating the mixing energy, one should consider the actual mixing equipment, fluid characteristics, and the design of the digester employed to determine the energy required rather than any general guidelines.

12.6 MISCELLANEOUS UTILITIES

Miscellaneous utilities are generally independent of the type of digestion system and the operational parameters of the digester. These include the building, heating energy, lighting energy, and ventilation energy.

Building heating energy depends on the ambient temperature, the volume of the building and number of air chutes, and is calculated by the energy required to maintain 20 °C inside the building. The lighting energy depends on the volume of the building and is calculated by the power required to produce an average of about a 215-lx lighting intensity. The ventilation energy, just like lighting energy, depends upon the volume of the building and is calculated by the power required to effect 5 volume changes of air per hour. Energy requirements for miscellaneous utilities are generally a small portion of the product energy.

12.7 SUMMARY

Net energy, which mainly depends on the total solids content of the feed and the operational parameters, must be considered in order to assess any digestion system for the conversion of organic matter to energy products. Substantial differences in the TS content among different biomass and organic wastes have significant influence on the net energy of any digestion scheme employed for their conversion. Feed heating energy is overwhelmingly the major portion of the total energy requirement, and for dilute feeds it can even exceed the total product energy. For such dilute feeds, digester operation at ambient temperature may be needed to reduce energy requirements. On the other hand, feeds with high solids content have very low feed heating energy requirements and can be digested at thermophilic temperatures and at short retention times without significantly affecting the net energy. It should be noted, however, that the digester design does not affect the feed heating requirement. In general, the feed heating requirement can be minimized by preheating the feed through heat exchange between the digester effluent and the feed slurry.

The net energy produced by a digestion system is also influenced by the loading rate. For any particular feedstock, the organic loading rate determines the digester size, which, in turn, affects the heat loss from the digester. Similarly, the product energy loss, that is, the heat loss in terms of methane, can be greater in high-flow-rate/low-retention-time digesters because the effluent leaving the digester may contain dissolved methane at about 50% or more of the saturation level.[12] In fact, the heat loss from a thermophilic digester operated at a higher loading rate can be the same as or lower than that from a mesophilic digester operated at a lower loading rate. The heat loss, in general, can be minimized by proper insulation and by locating digesters next to each other, if more than one digester is required for total operation. However, the effect of digester design, digester size, location, and mode of operation on energy loss should be evaluated further.

The digester design and the mode of operation also have an influence on the net energy, primarily because of the variations in energy losses and in pumping or mixing energy requirements. For example, the pumping energy requirement is greater in the fluidized-bed digester and upflow sludge blanket digester, due to continuous culture recirculation, than that in a conventional digester. On the other hand, some digester designs, such as upflow solids reactors and plug-flow reactors, require no mixing or additional pumping, thus resulting in slightly lower energy requirements. The pumping energy for some of the unconventional digestion systems such as fluidized- and expanded-bed digesters should therefore be calculated with regard to bed height, degree of fluidization, expansion, or recirculation, and the physical properties of the materials used for the bed. Also, the effect of rheological properties of different feedstocks on energy needs for pumping, mixing, etc., should be determined.

The effect of the heat of reaction generated during methane production has not been extensively studied. Limited information suggests that its effect is negligible and could be omitted in net energy calculations for low solid concentration feedstocks.[6,13] However, its effect on energy requirement for high solids feedstocks should be evaluated.

REFERENCES

1. *The Energy Conservation Paper*, R. H. Williams, Ed. Cambridge, Mass.: Ballinger Publishing Company, 1975.
2. Benemann, J. R., 'Biofuels: A Survey,' Special report prepared for Electric Power Research Institute, Palo Alto,Calif., June 1978.
3. Klass, D. L., 'Energy From Biomass and Wastes: 1982 Update,' Paper presented at *Energy From Biomass and Wastes VIII*, Institute of Gas Technology, Orlando, Fla, January 1983.
4. Fannin, K. F., Srivastava, V. J., Chynoweth, D. P. and Bird, K., 'Effects of the Interaction Between Biomass Composition and Reactor Design on Anaerobic Digester Performance,' Paper presented at *Energy From Biomass and Wastes VIII*, Institute of Gas Technology, Orlando, Fla, January 1983.
5. Ghosh, S., 'Net Energy Production in Anaerobic Digestion,' Paper presented at *Energy From Biomass and Wastes V*, Institute of Gas Technology, Orlando, Fla, January 1981.
6. Chen, Y. R. and Hashimoto, A. G., 'Energy Requirement for Anaerobic Fermentation of Livestock Waste,' ASAE Proceedings, *Livestock Wastes: A Renewable Resource*. St Joseph, Mich., 1981.
7. Kreider, J. F. and Kreith, F., *Solar Heating and Cooling — Active and Positive Design*, 2nd edn. New York: Hemisphere Publishing Corporation, 1982.
8. Stafford, D. A., Hawkes, D. L. and Horton, R., *Methane Production From Waste Organic Matter*. Boca Raton, Fla: CRC Press, 1980.
9. Chynoweth, D. P., Srivastava, V. J., Iverson, A. J. and Razik, A., 'A Research Study to Determine the Feasibility of Producing Methane Gas From Sea Kelp,' First Quarterly Report prepared for General Electric Company, IGT Project No. *30502*. Chicago: Institute of Gas Technology, April 1980.
10. Biljetina, R., Chynoweth, D. P., Janulis, J. and Srivastava, V. J., 'Biogasification of Walt Disney World Biomass Waste Blend,' Annual Report prepared for Gas Research Institute, IGT Project No. *30556*. Chicago: Institute of Gas Technology, May 1983.
11. Jewell, W. J., 'Anaerobic Attached Film Expanded-Bed fundamentals,' Proceedings of *First International Conference on Fixed Film Biological Processes*, 17–42, Kings Island, Ohio, April 1982.
12. Donaldson, T. L., Genung, R. K., Harris, M. T. and Rivera, A. L., 'Operation of an Upflow Fixed-Bed Anaerobic Digester for Waste Stabilization and Fuel Gas Production at Near-Commercial Scale,' Paper presented at

Energy From Biomass and Wastes VIII, Institute of Gas Technology, Orlando, Fla, January 1983.

13. Hashimoto, A. G., Chen, Y. R., Varel, V. H. and Prior, R. L., 'Anaerobic Fermentation of Agricultural Residues,' in *Utilization and Recycle of Agricultural Wastes and Residues*. Boca Rato, Fla: CRC Press, 1980.

13
Commercialization and Economics

RICHARD BILJETINA
Institute of Gas Technology, Chicago, Illinois, USA

13.1 INTRODUCTION

The ultimate goal of the research and development efforts on anaerobic digestion of biomass feedstocks is the production of a product fuel that is competitive with alternative sources. This chapter discusses the availability of applicable commercial data and information, scale-up criteria, and specific factors that will determine the economics of commercial digestion schemes. Since large-scale biomass digesters are yet to be placed in operation, data and information were limited to commercial or near-commercial applications using similar feedstocks. Extensive commercial experience exists for systems treating manure, sewage, food, and chemical wastes.

The design engineer should, therefore, first review the data from these commercial applications and then select digester designs that are best suited for the physical and chemical properties of the biomass and that will meet the performance and quality requirements dictated by the process scheme. Scale-up of the process should follow a logical progression of experimental work in laboratory- and pilot-scale equipment before evaluation in prototype or commercial-sized equipment. A summary of pertinent data ranging from laboratory to commercial scale is presented below.

13.2 COMMERCIAL ANAEROBIC DIGESTERS

This section summarizes some of the recent applications of anaerobic digesters in commercial or industrial waste treatment operations. The information is presented by major reactor type including batch, continuously stirred tank (CSTR), anaerobic filter, fluidized-bed, plug-flow, and upflow anaerobic sludge blanket.

Batch reactors were the first reactors used for anaerobic digestion of animal and human wastes.[1] Numerous installations still exist today, although many are small scale. Recent reports estimate 5–6 million batch reactors operating in China, 1 million in India, 30 000 in South Korea, 2300 in Brazil, and 1200 in Nepal.[2] The application of batch digestion in the large-scale processing of wastes has been limited because of relatively unstable and uncontrollable operation experienced in these systems. A variation of the batch digestion scheme, landfills with methane recovery, is providing data and operating experience that may be readily applied to biomass processing. Table 13.1 lists some recent applications. Investigations are being conducted to evaluate controlled landfills for biomass conversion.[6] Another batch process receiving considerable attention is the concept of dry fermentation[7,8] for digestion of agricultural residues.

Extensive experience also exists for commercial operations using the CSTR and stirred tank with solids recycle (CSTR/SR) technology. In addition to worldwide application in municipal sludge treatment, this technology has been used to treat effectively industrial wastes, manure, and municipal solid wastes (MSW). Table 13.2 lists some recent commercial applications for MSW, cattle manure, and sugar waste treatment.

Since sludge applications number in the thousands, we list only one as a typical example.[17] We should note, however, that operating data are often not readily available and that these digesters are operated at long residence times to ensure reduction of wastes. Gas production and reactor size are not optimized.

Table 13.3 lists full-scale applications of the anaerobic filter process. All commercial-scale operations are for the treatment of wastewater from industrial sources. The Celanese Chemical Company is employing the upflow anaerobic filter process to treat wastewater from a guar gum plant and two chemical plants. Successful operation was reported in a recent industrial wastewater working session for the Pampa and Bishop plants.[21] The Bacardi Corporation is using the anaerobic filter process in the downflow mode at their rum distillery plant in San Juan, Puerto Rico.[18,19,22] Operation began initially in December 1981. The digesters are now approaching design performance after a start-up period covering approximately 1.5 years.

TABLE 13.1
Landfill (Batch) Systems

Description		Gas product			
Location	*Capacity, hectares*	*Gas produced, $10^6\ m^3/day$*	*% CH_4*	*Remarks*	*Reference*
Reserve Synthetic Fuels Inc., Palos Verdes, California	16·19	0·057	NA*		
Duarte, California	13·35 (18·29 meters depth)	0·028	NA	Landfill to methane to electricity	4
Fresh Kills Landfill, New York City	16·19	0·28	NA		5
Operating & Ind. Landfill, Monterey Park, California	60·7 (18·15 million tonnes of MSW)	0·22	NA		3

*NA means not available

TABLE 13.2
Continuously Stirred Tank Anaerobic Digesters

Digester description				Operating data				Gas product			
Feed stock	*Process developer*	*Location*	*Capacity*	*HRT, days*	*Feed COD, mg/liter*	*COD load, kg/m³-day*	*COD removed, %*	*Gas produced*	*% CH_4*	*Remarks*	*Reference*
Municipal solid wastes	Waste Management, Inc.	Refcom, Pompano Beach, Fla	90·74 tonnes/ day at thermophilic, 45·37 tonnes/ day at mesophilic	10	NA	NA	NA	0·44–0·5 standard m^3 biogas/kg VS fed	50–55		9, 10, 11
Cattle manure	Thermonetics	Guymon, Okla.	544·46 tonnes/day (wet)	12	NA	NA	NA	64·2–77·08 m^3/tonne	65	Protein residues used as animal feed	9, 11, 12
Cattle manure	Biogas of Colorado	Leefer Farm, Carlinville, Ill.	567·75 m^3	NA	NA	NA	NA	1 217·8 m^3-day	60	Produced gas is used for heating water and structures, drying corn and fueling a fermentation alcohol plant	13
Cattle manure	Hamilton Standard, Inc.	Bartow, Fla	22·69 tonnes/day (dry)	10–40	101 920–117 120	1·12–5·61	NA	0·19–0·31 m^3 CH_4/kg VS fed	60	Produced gas supplies energy requirements of the associated meat packing plant boilers	14, 15
Sugar wastes	AB (Anamet Process) Sorigona	American Crystal Sugar Co., East Grand Forks, Minn.	34·48 tonnes BOD_5/day	NA	NA	NA	97·5 BOD Red.	NA	NA	Additional plants: Ortofta sugar factory, Sweden, Pfeifer & Langen's sugar factory, Euskirchen, Germany	16
Sludge	NA*	MSD, Chicago, Ill.	NA	17	56 900	2·08	27	56 640 m^3/day	65		Verbal communication

*NA means not available.

TABLE 13.3
Anaerobic Filter Digesters

Digester description					Operating data				Gas product			
Feed stock	*Process developer*	*Location*	*Capacity, m^3/day*	*Medium*	*HRT, days*	*Feed COD, mg/liter*	*COD load, kg/m^3-day*	*COD removed %*	*Gas produced*	*% CH_4*	*Remarks*	*Reference*
Rum distillery	Bacardi Corp.	San Juan, Puerto Rico	1 514–1 703	B. F. Goodrich Synthetic 'Vinyl Core'	8–10	85 000–100 000	8·81	65–70	0·53 m^3/kg COD rmvd	50–55		9, 18, 19, 20
Guar gum	Celanese Chemical Co.	Vernon, Tex.	823	9 cm Pall rings	1	9 140	16·02	60	0·37 m^3/kg COD rmvd	NA*		21, 22, 23
Chemical	Celanese Chemical Co.	Bishop, Tex.	3784	9 cm Pall rings	1·5	12 000	9·61	80	8 015 m^3/day	70	Product gas burned in boilers to produce steam	21, 22
Chemical	Celanese Chemical Co.	Pampa, Tex.	3784	9 cm Pall rings	1·5	14 400	10·41	90	0·36 m^3/kg COD rmvd	NA*		21, 22
Confectionery waste	Biomass Internat'l	NA*	NA*	NA*	NA*	14 800 (2 000–48 000)	NA*	97	0·49 m^3/kg COD apld	71		24
Molasses distillery waste	Biomass Internat'l	NA*	NA*	NA*	NA*	70 000	NA*	97	0·54 m^3/kg COD apld	64		24
Edible oil waste	Biomass Internat'l	NA*	NA*	NA*	NA*	2 620	NA*	81	0·55 m^3/kg COD apld	61		24

*NA means not available.

Commercial-scale applications of fluidized-bed reactors are listed in Table 13.4. The Ecolotrol unit in Birmingham, Alabama, is no longer operating, due to a long and difficult start-up period.[25] The Dorr Oliver installations were first tested on a pilot plant scale.[27]

Commercial plug-flow reactors are primarily used in the agricultural industry to treat dairy cow and poultry wastes. Table 13.5 lists several applications. Energy Cycle, Inc., actively markets plug-flow digesters in the United States, and Abetong-Sabema concentrates similar efforts in Europe. A very simplified form of plug-flow digestion was erected at a 1000 sow, hog farm in California, where a lagoon has been partially enclosed with a hypalon cover.[31] Approximately 0·4 hect, representing one-third of the lagoon, is covered and produces between 708 and 991 m^3 of biogas daily at lagoon temperatures between 10 and 22 °C. The biogas fuels an electric generator and a payback period of 2·9 years is reported.

Table 13.6 gives commercial applications of upflow anaerobic sludge blanket (UASB) digesters. Most of these facilities are located in Europe and include installations for the digestion of sugar beet, potato, starch, alcohol, and sugar wastes. Installations in the United States include one in Maine and two in Wisconsin. These full-scale digesters are treating wastes from potato-, starch-, and alcohol-processing facilities.

Although these tables indicate the abundance of information and experience available for different types of digester designs in commercial operation today, it should be recognized that the important design and operating data are not readily available to the scientific community. Digester and operating information is frequently proprietary and protected by secrecy agreements. Consequently, 'development efforts are fragmented; no consensus has yet emerged on design, operating conditions, appropriate end uses, feed/design matches and associated cost/benefit questions ... because of a lack of solid operating data from the bewildering array of existing and planned systems.'[2]

Since dissemination of operating data from commercial anaerobic digestion systems is limited, it becomes even more important to use proper scale-up techniques in the development of new biomass digestion processes.

13.3 SCALE-UP OF BIOMASS DIGESTERS

Using the design criteria discussed in this report and as much operating data as are available from similar commercial operations, a biomass

TABLE 13.4
Fluidized-Bed Digesters

Digester description					Operating data				Gas product			
Feed stock	*Process developer*	*Location*	*Capacity, m^3/day*	*Medium*	*HRT, h*	*Feed COD, mg/liter*	*COD load, kg/m^3-day*	*COD removed %*	*CH_4 gas produced*	*% CH_4*	*Remarks*	*Reference*
Soft drink bottling waste	Ecolotrol	Buffalo Rock Bottling (Pepsi), Birmingham, Ala.	378·5	0·6 mm ES sand	6	6900	9·61	77	NA*	NA*	No longer in operation	22, 25
Soy processing waste	Dorr-Oliver (Anitron)	Iowa	NA*	0·7 mm quartzite sand	11·8–20·2 <24	9000	12·98	NA*	NA*	70–75	Gas produced is used in an existing natural gas-fired operation	22, 26
Dairy (whey permeate)	Dorr-Oliver (Anitron)	Stamford, Conn.	NA*	0·7 mm 0·5 mm sand	12·5	8423	16·02	63–68	0·27 m^3/kg COD rmvd	60	Pilot plant	27

*NA means not available.

TABLE 13.5
Plug Flow Digesters

Digester description				*Operating data*				*Gas product*			
Feed stock	*Process developer*	*Location*	*Capacity*	*HRT, days*	*Feed COD*	*COD load, kg/m³-day*	*COD removed*	*Gas produced*	*% CH_4*	*Remarks*	*Reference*
Dairy cow manure	Harvest Co. of Chicago	Fertile Acres Farm, Rice Lake, Wisc.	3·63 tonnes/day (wet)	15	NA*	NA*	NA*	31·2 m³/ tonne	NA*	Produced gas drives 12 kW engine generator set	9
Dairy cow manure	Energy Harvest	Mason Dixon Farm, Gettysburg, Pa	50·98 m³/day	13	NA*	NA*	NA*	990 m³/ day	60	Gas is used to produce electricity. Effluent used for bedding	12, 28
Dairy cow manure	Energy Cycle	Baum Dairy, Springport, Mich.	681·3-m³ unit	10–15	9·3 wt % solids	NA*	NA*	920 m³/ day	20 492 kJ/m³	Produced gas used for farm's electrical needs	29, 30
Dairy waste/ cheese whey	Energy Cycle	R&M Farms, Hobart, NY	908·4-m³ unit	NA*	NA*	NA*	NA*	NA*	55–60	Power sold to local utility	29
Dairy waste	Energy Cycle	Sunny Valley Farms, New Milford, Conn.	227·1-m³ unit	NA*	NA*	NA*	NA*	NA*	55–60	Produced gas supplies parallel service with power company	29
Poultry waste	Energy Cycle	Warren Falls, Melbourne, Alaska	340·7-m³ unit	NA*	NA*	NA*	NA*	425 m³/ day	55–60	Produced gas supplies parallel service with power company	29
Dairy cow	Energy Cycle	Circle B Farms, Orland, Calif.	908·4-m³ unit	NA*	NA*	NA*	NA*	NA*	55–60	Produced gas provides electricity for milking operation	29

*NA means not available.

treatment scheme can be selected for initial testing in laboratory-scale equipment. Such tests can provide initial data on reaction rates, methane yields, and residue composition. These data serve as a basis for scale-up to a pilot plant size that is sufficiently large so that information on materials handling, solids loading, energy requirements, and physical reactor and equipment designs can be obtained to establish process configurations and preliminary economics. The importance of evaluating and solving materials-handling problems at the pilot scale cannot be overemphasized. Costly materials-handling problems have been experienced in prototype commercial digesters designed solely on the basis of laboratory experience without the benefit of intermediate-scale evaluation. Therefore, only after successful completion of these phases should prototype or commercial-sized plants be constructed.

An example of such a progression of scale-up is the work being conducted at IGT on an upflow solids reactor (USR) digestion system for conversion of biomass and biomass/sludge fractions to methane.[34] After study of available digester systems, a nonstirred, USR design was chosen. Laboratory work was initiated in 5-liter digesters to study digestion of kelp[33] and water hyacinth.[34] Laboratory results showed that high solids retention times could be achieved in this digester and that, at conventional loading rates, gas production was about 20% highér than in CSTR designs.

To further the development of this digestion scheme, a 4·53-m^3 digester has been constructed to treat water hyacinth/sludge mixtures. The primary function of this unit is to validate baseline digester operation and performance obtained in the laboratory units and to evaluate larger-scale equipment for chopping, slurry preparation, mixing, pre- and post-treatment, and dewatering. This unit will also provide sufficient effluent for evaluation of its potential animal feed and fertilizer value. The results from these studies will provide a basis for a complete conceptual process design and preliminary cost estimate of a larger conversion sytem.

Similar efforts are also in progress for the development of other digester schemes. Tables 13.7 and 13.8 list some laboratory and pilot plant efforts on two-phase anaerobic digestion. Application of this two-phase process is increasing because of its high degree of stability. Numerous other concepts and techniques are also under study.[39–41] Biomass-based anaerobic digesters for commercial application will be successfully developed when sufficient data are gathered in laboratory- and pilot-scale digesters to support the engineering of the process concept and the determination of reliable cost projections and economics.

TABLE 13.6
Upflow Anaerobic Sludge Blanket (UASB) Digesters

Digester description				Operating data				Gas product			
Feed stock	*Process developer*	*Location*	*Capacity*	*HRT, h*	*Feed COD, mg/liter*	*COD load, kg/m³-day*	*COD removed, %*	*Gas produced, m³/h*	*% CH_4*	*Remarks*	*Reference*
Brewery	Joseph Oats Corp.	Heilemann Brewery, LaCrosse, Wisc.	961·4 m³/h	4·9	2 500	16·34	80	35 300	75	Design Data	22, 30
Potato-starch	Joseph Oats Corp.	Colby, Caribou, Me.	37·85 m³/h	47	22 000	11·05	85	7 800	77	Design Data	22, 30
Potato	Joseph Oats Corp. (Biomethane)	Plover, Wisc.	124·9 m³/h	17·5	4 300	5·93	80	7 100	70	Design Data	22, 32
Sugar beet	NA*	Netherlands	NA*	24	17 000	13·30	94	NA*	NA*	Design Data	22
Sugar beet	NA*	SU, Netherlands	20 657 kg COD/day	4·9	3 000	14·42	85	NA*	NA*	Design Data	22, 30
Potato-starch	NA*	Avebe, Netherlands	13 510 kg COD/day	24	7 700	8·01	85	NA*	NA*	Design Data	22, 30
Potato	NA*	Cisa, Switzerland	4 990 kg COD/day	6·5	2 260	8·33	80	NA*	NA*	Design Data	22, 30
Sugar beet	NA*	Bruhl, Germany	17 963 kg COD/day	15	7 500	12·02	86	NA*	NA*	Design Data	22, 30

Alcohol	NA*	ZNSF, Netherlands	10 977 kg COD/day	8	5 330	16·02	90	NA*	NA*	Design Data	22, 30
Sugar beet	CSM	CSM Halfweg, Netherlands	9 580 kg COD/day	4	1 850	12·02	70	120	82	Operational Results	30
Sugar beet	CSM	CSM Breda, Netherlands	15 568 kg COD/day	4·8	2 400	12·02	75	200	82	Operational Results	30
Sugar beet	CSM	SU Groningen, Netherlands	23 464 kg COD/day	5·7	4 000	16·02	75	300	76	Operational Results	30
Sugar beet	NA*	Sugana, Austria	24 948 kg COD/day	NA*	NA*	8·17	NA*	NA*	NA*	Design Data	30
Sugar beet	NA*	SU, Netherlands	27 942 kg COD/day	NA*	NA*	15·54	NA*	NA*	NA*	Design Data	30
Potato	NA*	FF, Netherlands	1 597 kg COD/day	NA*	NA*	6·73	NA*	NA*	NA*	Design Data	30
Potato	NA*	Aviko, Netherlands	3 193 kg COD/day	NA*	NA*	8·01	NA*	NA*	NA*	Design Data	30
Potato	NA*	Aviko, Netherlands	15 967 kg COD/day	NA*	NA*	10·57	NA*	NA*	NA*	Design Data	30
Sugar beet	NA*	CSM, Netherlands	28 940 kg COD/day	NA*	NA*	17·62	NA*	NA*	NA*	Design Data	30

*NA means not available.

TABLE 13.7
Two-Phase Anaerobic Digestion (Laboratory)

Digester description			*Operating data*				*Gas product*			
Feed stock	*Process developer*	*Location*	*HRT, days*	*Feed COD*	*COD load, kg/m³-day*	*COD removed, %*	*Gas produced*	*% CH_4*	*Remarks*	*Reference*
Straw (roughly chopped)	University College, Dept. of Microbiol.	Galway, Ireland	18	NA*	NA*	NA*	0·19 m³/kg TS	85		35
Straw (NaOH treated)	University College, Dept. of Microbiol.	Galway, Ireland	18	NA*	NA*	NA*	0·30 m³/kg TS	85		35
Grass (clippings)	University College, Dept. of Microbiol.	Galway, Ireland	8–14	NA*	NA*	NA*	0·28 m³/kg TS	85		35
Soft drink waste (South Atlantic State)	IGT	Chicago, Illinois	7·4	NA*	4·81	96	0·61 standard m³/kg VS added	63·1	Acid phase: 2·5 liter completely mixed digester. Methanogenic phase: 5·5 liter upflow anaerobic filter	36

*NA means not available.

TABLE 13.8
Two-Phase Anaerobic Digestion (Pilot Plant and Full-Scale)

Digester description				*Operating data*				*Gas product*			
Feed stock	*Process developer*	*Location*	*Capacity*	*HRT, days*	*Feed COD, mg/liter*	*COD load, kg/m³-day*	*COD removed, %*	*Gas produced*	*% CH_4*	*Remarks*	*Reference*
Distillery Works, D.C.M. Sugar Division	Biochemical Engr. Research Center, Indian Inst. of Tech.	Wastes from North-central India	Bench scale (2, 10, 20, & 40 liter)	10	NA*	35·24	73	30 liter CH_4/ liter waste	NA*		37
Distillery Works, E.I.D. Parry Ltd.	Biochemical Engr. Research Center, Indian Inst. of Tech.	Wastes from South India	Pilot scale (250 & 1000 liter)	10	NA*	35·24	73	30 liter CH_4/ liter waste	NA*		37
Distillery Works, Saktai Sugars Ltd.	Biochemical Engr. Research Center, Indiän Inst. of Tech.	Wastes from South-central India	Pilot scale (250 & 1000 liter)	10	NA*	35·24	73	30 liter CH_4/ liter waste	NA*		37
Yeast and alcohol	Studiebureau O. Dekonickx	Belgium	Pilot scale	2·8	32 000	11·70	72	8·4 liter gas/liter reactor-day	NA*		38
Beet sugar	Studiebureau O. Dekonickx	Belgium	Pilot scale	0·6	6 000	10·90	90	7·2 liter gas/liter reactor	NA*		38
Flax retting	Studiebureau O. Dekonickx	Belgium	Full scale	0·7	7 000	10·10	87	4·2 liter gas/liter reactor-day	NA*		38

*NA means not available.

Economics, of course, must be the final decision factor in the commercialization of a process.

13.4 ECONOMICS

The economics of methane production by anaerobic conversion of biomass is affected by a number of factors determined not only by the selected process scheme but also by the availability of suitable financial conditions. These factors will include the size of the operation, the location and site requirements, the feedstock used, the digestion concept, product utilization and credit allowances, environmental constraints and surcharges in effect, market rates, tax incentives, the cost and availability of money, and projected competition from alternative energy sources. It is important that all of these factors be considered in the determination of projected rates of return for any project.

Because so many factors are involved in a meaningful cost analysis, it is not possible to generalize the requirements for favorable economics. Each project has to be analyzed individually and scrutinized for its economic value. The potential economic benefits of anaerobic digestion are, however, evident from the attractive payback periods, rates of return, and gas costs reported in the literature for plants in operation today.

One of the simplified and commonly used calculations to indicate economic feasibility is the payback period. Some very short payback periods are being reported for farmer-owned digestion systems. The Leefer farm, which operates a conventional 567 750-liter digester on a 1000-head beef cattle manure influent, reports a payback period of 2·8 years.[13] A covered lagoon biogas system at a 1000 to 1100 sow farrow-to-finish hog farm in California reported a 2·9-year payback period (and an internal rate of return of 34%).[31]

Foreign experiences in similar applications also look promising. Anaerobic digestion of pig and cattle manure in large-scale digesters is reported to have a payback period of about 9 years in France and about 6·5 years in Singapore.[42] In addition, the payback period of a 5000 m^3 anaerobic contact digester operating on vegetable canning wastes, in France, is projected to have a payback period of 5–6 years.[43]

A number of system studies also have been made for farm applications, which relate size of the operation to economic feasibility. A 1978 study for the Office of Technology Assessment suggests that 49% of the manure produced annually in the United States from livestock operations is generated in large enough operations that the manure could be processed in digesters with economically favorable results. This could be increased to 70% if digester effluent is processed for cattle re-feed with a credit of $99·2/tonne dry weight or for enriched fertilizer at $11·0/tonne weight credit over the influent value. Additional benefits are also realized if owner installation and operation is contemplated, mainly because of the lower projected labor costs.[9] Similarly, Hashimoto and Chen[20] projected economic feasibility for beef feedlots using anaerobic digesters on 1000- to 2000-head operations without effluent credit and as small as 300 head for systems taking advantage of effluent feed credit. Once again, this assumed farmer-constructed and -operated units would have a lower breakeven cost than commercial 'turn-key' installations. A study by Martin and Loehr also indicates economic feasibility for manure operations on large dairy farms even at 20% annual returns. However, their studies showed that the economic breakeven point is very sensitive to the effectiveness of biogas utilization, the capital requirements, and the expected rate of return.[44]

When looking toward biomass conversion for the primary purpose of methane production, the projected gas cost is a more meaningful indicator of economic feasibility. Data from present operations in industrial and farm applications suggest favorable gas costs can be achieved. The Bacardi anaerobic filter, treating 1 892 500–2 573 800 liters/day of mosto wastes from its rum distillery, began operation in 1981 and is now projecting a gas cost of $1·42 to $1·90 per GJ.[18,19] A farm operation in Israel processing cattle manure in two 100 000-liter thermophilic digesters is reporting gas cost as low as $1·90 per GJ when 100% re-feed credit is applied.[45]

Currently, however, commercial data do not exist for large-scale biomass digestion. Therefore, information on the potential economic benefits of large-scale biomass digestion is limited to studies such as conducted on kelp and water hyacinth by the Gas Research Institute.[46,47] Projections include a detailed analysis of the cost components of the planned system. A fully integrated biomass conversion process is likely to contain five main areas:

(1) Planting, harvesting, transportation, and storage.
(2) Feed preparation and pretreatment.
(3) Anaerobic digestion.
(4) Gas clean-up and utilization.
(5) Digester effluent storage, treatment, and disposal.

TABLE 13.9
Basic Cost Elements of Gas Cost Calculations for a Biomass-Fed Anaerobic Digestion Plant

Capital Costs
Equipment—installed
Land
Start-up
Operating Costs
Supplies: feedstock, chemicals, additives, nutrients, water
Energy consumption: electricity, fuel
Materials: spare parts
Disposal: digester cake
Labor: operation, maintenance, supervision, overhead
Taxes and insurance
Credits: reduction in municipal surcharges, digester cake, tipping charges

Each section will have to be evaluated for capital and operating cost contributions. Table 13.9 lists some of the expected major cost contributors.

These costs must then be applied in a formula that includes applicable project and financial data (such as plant life, tax rates, tax credits, method of depreciation, debt-to-equity ratio, interest rate, inflation rate, and rate of return) to determine a gas cost and the economic feasibility of the project. Dynatech R/D Company has developed and modeled such a cost formula, for instance, for fuel gas production from animal residue.[48]

The project and financial data will generally be determined by the location and financial condition at the time of project initiation and, therefore, cannot be controlled by the design engineer. (It is important to note that the project and financial data should be on the same basis to arrive at meaningful conclusions about the merit of one project over another.) The cost elements of a project, however, can be controlled and optimized. Therefore, a more thorough discussion of these elements for each of the five main areas is included.

13.4.1 Planting, Harvesting, Transportation, and Storage

The planting and harvesting of biomass feedstocks for a digestion process represents potentially the largest additional cost contribution in both labor and land resource charges when compared to present commercial ventures in anaerobic digestion, where the feedstock is normally available at no charge to the process. In a recent study completed by Ralph M. Parsons Co. on large-scale (70 790 m^3/day of salable, pipeline-quality gas) methane production from kelp, the planting and harvesting section represented 45% of the total capital cost contribution and 60% of the total operating cost contribution. Feedstock selection and the size of the operation are, therefore, important criteria in the planning of a biogas project.

A number of alternatives have been suggested to minimize the impact of the land resource requirements. One is the use of our oceans and waterways to produce aquatic biomass feeds. For instance, to supply 100% of our national gas demand would require a land area for biomass production of 89 million hectares; however, this would represent only 0·2% of the available world ocean surface.[49] Approximately 235 million hectares alone are estimated to be available for marine biomass growth around US coastal shores.[50] Another method is to utilize our idle crop land. It is estimated that about 12% of our total US energy consumption could be met by harvesting biomass energy crops on idle land.[9] Another option is the collection of biomass residues from present operations. These resources are placed at 91 million dry tonnes with a theoretical maximum of 387 million dry tonnes.[9] A recent study by the Energy Technologies Group of Los Alamos National Laboratory placed the following collection costs in 1981 dollars (collection over a 16·1-kilometer radius):[51]

Corn stover	$21·63/tonne
Wheat straw	$24·93/tonne
Soybean residue	$37·80/tonne
Wood chips	$61·91/tonne

Jewell *et al.* estimated corn stover collection costs for a similar 16.1-kilometer radius between $45·2/tonne (worst case) to as low as $3·3/tonne, a very attractive feedstock cost.[52]

It is likely that aquatic plants and some grasses will make the best feedstocks for biogas production. These feedstocks are highly biodegradable (maximum efficiency), have a high water content (minimum water consumption), have low preprocessing requirements, have concentrated availability, and exhibit no toxicity to microorganisms. Forest residues, on the other hand, tend to be dry and contain significant amounts of lignin and have higher preprocessing costs.[9] Another factor that impacts on biomass selection and affects storage and plant operating costs is the availability and quality of certain feedstocks on a seasonal basis. Optimal economic results favor continuous operations at or near design rates.

The equipment necessary to harvest, transport, and store biomass feedstock should be commercially available. An exception may be large aquatic operations where development of special harvesting machinery may be desirable and result in lower capital and operating costs. The selection of biomass feedstocks is likely to have a significant effect on the capital and operating costs required to harvest, transport, and store the digester feed. It suggested that this area requires careful consideration of land requirements, feedstock properties, availability, and harvesting techniques to minimize the impact on the gas cost.

13.4.2 Feed Preparation and Pretreatment

The feed preparation area for biomass feedstocks is likely to include chopping and grinding equipment for size reduction. Classification equipment to separate undesirable fractions or contaminants may also be necessary. Transport equipment such as conveyor belts, drag conveyors, screw conveyors, or slurry pumps will be required to move the feedstock from station to station.

Pretreatment of the feedstock may be necessary to break down the lignocellulose fraction and enhance solubilization of the solids. This can be accomplished by physical treatment (particle size reduction or electron irradiation), by thermal treatment (direct and indirect heat, steam, pressure), or by chemical/biological treatment (acid, base, enzymes).[9]

The extent of feed preparation and pretreatment necessary will determine the additional capital, energy, supplies, and labor cost required for a particular feedstock. For the most part this equipment should be commercially available. Two exceptions may occur:

(1) Large energy plants could benefit from scale-up of choppers or grinders to larger-than-normal production rates.
(2) The selection of innovative pretreatment processes may require development.

13.4.3 Anaerobic Digestion

As discussed earlier the selection of digester type is likely to be determined by the feedstock characteristics. It will be important to evaluate the effect of retention time, loading rate, and solids concentration in the digester on the tank size and, therefore, the capital cost.

Capital costs for the following equipment, as required, must be included:

- Digester tank and accessories (internals, gas separators, media, scum removers)
- Mixers
- Pumps
- Settling/separation devices (sedimentation tanks, degasifiers, membranes)
- Heating system and insulation
- Instruments and controls
- Housing
- pH and nutrient control equipment

Energy requirements to heat the influent and maintain digester temperatures must be considered. Evaluation of heat loss versus insulating costs and heat exchange between effluent and influent to affect energy savings should be considered. It has been reported that maintenance of system temperatures can consume between 10% and 60% of the energy produced.[9]

Operating options at psychrophilic (less than 20 °C), mesophilic (20–45 °C) and thermophilic (45–65 °C) temperatures may also be a consideration. The selection of a digestion process should be guided by the desire to combine simple construction, high loading rates, and energy conservation to achieve high conversion rates and gas yields.

13.4.4 Gas Cleanup and Utilization

The use of the product gas will determine, in most cases, the extent of gas cleanup and compression. A natural gas substitute fuel would have one of the higher requirements including moisture removal, carbon dioxide (CO_2) and hydrogen sulfide (H_2S) scrubbing, and compression to pressures as high as 6·9–10·3 MPa (gauge). Cogeneration schemes may be used to improve the overall thermal efficiency and to enhance economics.

Equipment for safely storing and flaring the gas may be included and, if on-site electric generation is desired, the power generation and distribution system must be included. Large facilities are also likely to require tighter compliance with environmental regulations for gas flaring and fugitive emissions.

Equipment, in general, is commercially available; however, detailed engineering analyses are required for the selection of CO_2 and H_2S removal systems. Larger plants may benefit from more sophisticated, high-pressure scrubbers by reducing capital costs.

13.4.5 Digester Effluent Treatment, Storage, and Disposal

Since potential benefits from the use of digester cake for fertilizer, animal food substitutes and soil conditioners are very significant (one experience in Israel showed that product gas costs could be reduced from $10·42 to $1·95 per GJ by selling the digester cake for animal re-feed),[45] it is recommended that such options be routinely evaluated for any concept. It is important to note that anaerobic digestion is especially suited to the preservation of nutrient value while producing a valuable byproduct gas from a pollutant source.[9] Therefore, a digestion process, besides producing a valuable energy source, produces effluent solids that may have market value and, in some cases, the process lowers the municipal surcharge to the owner because of the reduction in volatile solids and BOD content in the effluent finally discharged to a municipal treatment center.

On the other hand, biogas production facilities may have disposal costs associated with the final disposition of digester effluents. Equipment will include pumping and intermediate storage facilities. If solids recovery is desirable, settling basins, presses, vacuum filters, and solids transport devices will add to the capital costs.

13.5 SUMMARY

Currently there are no large-scale, anaerobic digesters in operation that process a biomass feedstock. Proper commercial development will depend on matching the physical and chemical characteristics of the feedstock with available digester types. A review of existing commercial technologies operating on similar feedstocks should help in the initial selection of a digester design. It will be important, however, to conduct sufficient research at the laboratory scale to establish first an adequate data base of expected reaction rates and product yields for the selected digestion system. This must then be followed by scale-up to a pilot-plant-sized facility where materials handling, physical configurations, and operating mechanics can be sufficiently determined to provide preliminary economic projections.

Final process configuration must then take into account product utilization and quality requirements of the market, as well as the appropriate financial factors to arrive at projected gas costs with reasonable confidence levels. Such economic assessments would benefit from continuing efforts to:

(1) Widen the available data base on the sensitivity of gas costs to biomass feedstock costs, harvesting costs, pretreatment costs, digester costs, gas clean-up costs and utilization costs.
(2) Identify markets that would benefit from the combination of gas production and effluent treatment for environmental and reuse purposes.
(3) Identify markets requiring minimal gas treatment and efficient gas usage.
(4) Identify biomass resources and complete system studies on large-scale harvesting techniques.

Such efforts combined with continuing efforts to identify improved digester designs coupled with proper development and documentation through the various scale-up stages should ensure that the commercialization of biomass-based anaerobic digestion systems is successful and economically sound.

REFERENCES

1. McCarty, P. L., 'One Hundred Years of Anaerobic Treatment,' *Anaerobic Digestion 1981*, Proceedings of the Second International Symposium on Anaerobic Digestion, Travemunde, Federal Republic of Germany, 6–11 September 1981.
2. 'Special Report: 3rd International Symposium on Anaerobic Digestion,' *Biomass Digest*, September 1983.
3. Rice, F. R., 'Second Generation Technology for the Production of Pipeline Quality Gas From Sanitary Landfills,' *Proceedings of Energy from Biomass and Wastes IV*. Chicago: Institute of Gas Technology 1980.
4. 'Emerging Technologies Profile,' *Biomass Digest*, May 1983.
5. Vaszily, J. A., 'Commercialization of Landfill Gas Recovery: Fresh Kills Case Study,' *Proceedings of Energy from Biomass and Wastes VII*. Chicago: Institute of Gas Technology, 1983.
6. Wise, D. L., *et al.*, '*In Situ* Methane Fermentation of Combined Agricultural Residues,' *Resources and Conservation*, Vol. **6**:275–294. Amsterdam: Elsevier Scientific Publishing Company, 1981.
7. Jewell, W. J., *et al.*, 'Agricultural and High Strength Wastes,' pp. 169–184 in *Anaerobic Digestion 1981*, Proceedings of the Second International Symposium on Anaerobic Digestion, Travemunde, Federal Republic of Germany, 6–11 September 1981.
8. Jewell, W. J., *et al.*, 'Low Cost Approach to Methane Generation, Storage and Utilization From Crop and Animal Residues,' Fourth Quarter Progress Report for Period from 1 October 1980 to 31 December 1980, prepared for Solar Energy Research Institute under Contract No. XB-9038-1.

9. Abeles, T., Ellsworth, D. and Genereux, J., *TaskVI, Biological Production of Gas for Office of Technology Assessment*, Washington, DC, April 1979.
10. Walter, D. K., 'Anaerobic Digestion of Municipal Solid Waste to Produce Methane,' *Proceedings of Energy from Biomass and Wastes VII*. Chicago: Institute of Gas Technology, 1983.
11. Klass, D. L., 'Energy From Biomass and Wastes: 1981 Update,' *Proceedings of Energy from Biomass and Wastes VI*. Chicago: Institute of Gas Technology, 1982.
12. 'New Processes Harvest Farm Energy,' *Chem. Week* **129**, 15:41, 7 October 1981.
13. Schellenbach, S., 'Case Study of a Farmer Owned and Operated 1000 Head Feedlot Anaerobic Digester,' pp. 545–562 in *Proceedings of Energy from Biomass and Wastes VI*. Chicago: Institute of Gas Technology, 1982.
14. Stafford, D. A., Hawkes, D. L. and Horton, R., *Methane Production from Waste Organic Matter*. Boca Raton, Fla: CRC Press, Inc., 1980.
15. Lizdas, D. J. and Coe, W. B., 'Operation of a Full-Scale, Integrated Manure Anaerobic Fermentation Facility,' *Proceedings of Energy from Biomass and Wastes V*. Chicago: Institute of Gas Technology, 1981.
16. Huss, L., 'Applications of the Anamet Process for Waste Water Treatment,' pp. 137–150 in *Anaerobic Digestion 1981*, Proceedings of the Second International Symposium on Anaerobic Digestion, Travemunde, Federal Republic of Germany, 6–11 September 1981.
17. Farrell, J. B., 'Recent Developments in Anaerobic Sludge Digestion in the United States and a View of the Future,' Paper presented at the Conference Stabilization and Disinfection of Sewage Sludge, University of Manchester, England, 12–15 April 1983.
18. Szendrey, L. M., Schafer, P. E. and Dorion, G. H., 'Pollution and Energy Management Through the Anaerobic Approach,' *Industrial Wastes*, 31–34, Sept./Oct. 1982.
19. Szendrey, L. M., 'The Bacardi Corporation Digestion Process for Stabilizing Rum Distillery Wastes and Producing Methane,' *Proceedings of Energy from Biomass and Wastes VII*. Chicago: Institute of Gas Technology, 1983.
20. Hashimoto, A. G. and Chen, Y. R., 'Economic Optimization of Anaerobic Fermenter Designs for Beef Production Units,' pp. 129–132 in *Livestock Waste: A Renewable Resource*, Proceedings of 4th International Symposium on Livestock Wastes—1980, Amarillo, Texas, 1981.
21. Sheridan, R. P., 'The Celrobic System—Failures and Successes,' Paper presented at the Symposium Anaerobic Biotechnology: Reducing the Cost of Industrial Wastewater Treatment, Arlington, Va, 14 April 1983.
22. Bowker, R., 'Anaerobic Systems for Industrial and Municipal Wastewater Treatment,' Paper presented at A.I.Ch.E. Meeting, Session 32, Cleveland, 29 August–1 September 1982.
23. Witt, E. R., Humphrey, W. J. and Roberts, T. E., 'Full Scale Anaerobic Filter Treats High Strength Wastes,' p. 229 in *Proceedings of the 34th Industrial Waste Conference*, Purdue University, 1979.
24. Anderson, G. K., Donnelly, T. and McKeown, K. J., 'Application of Anaerobic Packed-Bed Reactors to Industrial Wastewater Treatment,' pp.

651–660 in *Proceedings of the 37th Industrial Waste Conference*, Purdue University, May 1982.
25. Miller, W. D., 'Anaerobic Pretreatment and Soft Drinks,' Paper presented at the Symposium Anaerobic Biotechnology: Reducing the Cost of Industrial Wastewater Treatment, Arlington, Va, 14 April 1983.
26. Sutton, P. M. and Li, A., 'Anitron System and Oxitron System: High Rate Anaerobic and Aerobic Biological Treatment Systems for Industry,' pp. 665–677 in *Proceedings of the 36th Industrial Waste Conference*, Purdue University, May 1981.
27. Genung, R. K., Harris, M. T., Rivera, A. L. and Donaldson, T. L., 'Operation of an Upflow Fixed-Bed Anaerobic Digester for Waste Stabilization and Fuel Gas Production at Near-Commercial Scale,' *Proceedings of Energy from Biomass and Wastes VII*. Chicago: Institute of Gas Technology, 1983.
28. Martin, J. H. and Lichtenberger, P. L., 'Operation of a Commercial Farm-Scale Plug-Flow Manure Digester Plant,' *Proceedings of Energy from Biomass and Wastes V*. Chicago: Institute of Gas Technology, 1981.
29. Energy-Cycle, Inc., general information from brochures about existing anaerobic digester systems and projected systems.
30. Pette, K. C. and Versprille, A. I., 'Application of the U.A.S.B. — Concept for Wastewater Treatment,' pp. 121–133 in *Anaerobic Digestion 1981*, Proceedings of the Second International Symposium on Anaerobic Digestion, Travemunde, Federal Republic of Germany, 6–11 September 1981.
31. Chandler, J. A., Hermes, S. J. and Smith, K. D., 'A Low Cost 75 kW Covered Lagoon Biogas System,' *Proceedings Energy from Biomass and Wastes VII*. Chicago: Institute of Gas Technology, 1983.
32. Christensen, D. R., 'A Biothene Case Study — Orelda Foods,' Paper presented at the symposium on Anaerobic Biotechnology: Reducing the Cost of Industrial Wastewater Treatment, Arlington, Va, 14 April 1983.
33. Institute of Gas Technology, 'Marine Biomass Program: Anaerobic Digester Systems Development and Study,' GRI Contract No. 5802-225-0607, IGT Project No. 30558 Letter Report for November 1982.
34. Chynoweth, D. P., Dolenc, D. A., Schwegler, B. and Reddy, K. R., 'Wastewater Reclamation and Methane Production Using Water Hyacinth and Anaerobic Digestion,' Paper presented at 10th Energy Technology Conference, Washington, DC, 28 February–2 March 1983.
35. Colleran, E., Barry, M. and Wilkie, A., 'The Application of the Anaerobic Filter Design to Biogas Production from Solid and Liquid Agricultural Wastes,' *Proceedings of Energy from Biomass and Wastes VI*. Chicago: Institute of Gas Technology, 1982.
36. Ghosh, S. and Henry, M., 'Stabilization and Gasification of Soft-Drink Manufacturing Waste by Conventional and Two-Phase Anaerobic Digestion,' pp. 282–300 in *Proceedings of the 36th Industrial Waste Conference*, Purdue University, May 1981.
37. Das, D., Ghose, T. K., Joshi, A. P. and Gopalkrishnan, K. S., 'Treatment of Distillery Wastes by a Two Phase Biomethanation Process,' *Proceedings of Energy from Biomass and Wastes VII*. Chicago: Institute of Gas Technology, 1983.

38. Ghosh, S., Ombregt, J. P., DeProost, V. H. and Pipyn, P., 'Methane Production From Industrial Wastes by Two-Phase Anaerobic Digestion,' *Proceedings of Energy from Biomass and Wastes VI.* Chicago: Institute of Gas Technology, 1982.
39. Stafford, D. A. and Etheridge, S. P., 'Farm Wastes, Energy Production and the Economics of Farm Anaerobic Digesters,' pp. 255–268 in *Anaerobic Digestion 1981*, Proceedings of the Second International Symposium on Anaerobic Digestion, Travemunde, Federal Republic of Germany, 6–11 September 1981.
40. Hashimoto, A. G., 'Conversion of Straw–Manure Mixtures to Methane at Mesophilic and Thermophilic Temperatures,' *Biotechnol. and Bioeng.* **25**:185–200, 1983.
41. Hashimoto, A. G. and Robinson, S. A., 'Pilot-Scale Conversion of Manure–Straw Mixtures to Methane,' *Resources and Conservation*, Vol. **6**:19–28. Amsterdam: Elsevier Scientific Publishing Company, 1981.
42. Aubart, C. and Farinet, J. L., 'Anaerobic Digestion of Pig and Cattle Manure in Large Scale Digesters and Power Production From Biogas,' *Proceedings of Energy from Biomass and Wastes VII.* Chicago: Institute of Gas Technology, 1983.
43. Morfaux, J. N., Touzel, J. P. and Albagnac, G., 'Anaerobic Digestion of Vegetable Canning Wastewaters,' pp. 185–199 in *Anaerobic Digestion 1981*, Proceedings of the Second International Symposium of Anaerobic Digestion, Travemunde, Federal Republic of Germany, 6–11 September 1981.
44. Martin, J. H., Jr, and Loehr, R. C., 'Economic Analysis of Biogas Production and Utilization,' pp. 327–329 in *Livestock Wastes: A Renewable Resource*, Proceedings of 4th International Symposium on Livestock Wastes, 1980, Amarillo, Texas, 1981.
45. Marchaim, U., Prochenski, D., Perach, Z., Lombrozo, E., Dvoskin, D. and Criden, J., 'The Israeli Anaerobic Digestion Process — NEFAH,' pp. 203–216 in *Anaerobic Digestion 1981*, Proceedings of the Second International Symposium of Anaerobic Digestion, Travemunde, Federal Republic of Germany, 6–11 September 1981.
46. Brehany, J. J., 'An Economic and Systems Assessment of the Concept of Nearshore Kelp Farming for Methane Production,' Final Report from June 1982 to April 1983, prepared for Gas Research Institute under Contract No. 5082-511-0627, 1983.
47. Black & Veatch, 'Economic and Systems Assessment of the Concept of a Water Hyacinth Wastewater/Methane Production System,' Final Report from May 1982 to May 1983, prepared for Gas Research Institute under Contract No. 5082-511-0629, 1983.
48. Ashare, E., Wentworth, R. L. and Wise, D. L., 'Fuel Gas Production from Animal Residue Part II. An Economic Assessment,' *Resource Recovery and Conservation* Vol. **3**:359–386. Amsterdam: Elsevier Scientific Publishing Co., 1979.
49. Wise, D. L., Wentworth, R. L. and Ashare, E., 'Methane Production from Biomass and Agricultural Residues,' *Ind. Eng. Chem. Prod. Res. Dev.*, Vol. **18**, No. 2, 1979.

50. Snow, J. T., *et al.*, Proceedings of EPRI/GRI Workshop on Biomass Resources and Conversion, Santa Clara, California, 1978.
51. Thayer, G. R., Rollett, H., Rood, P. L. and Williamson, K. D., Jr, 'Levelized Life Cycle Costs for Four Residue Collection Systems and Four Gas Production Systems,' *Proceedings of Energy from Biomass and Wastes VII*. Chicago: Institute of Gas Technology, 1983.
52. Jewell, W. J., Cummings, R. J., Dell'Orto, S., Fanfoni, K. J., Fast, S., Gottung, E. J., Jackson, D. and Kabrick, R. M., 'Dry Fermentation of Agricultural Residues,' Final Report No. SERI/STR-231-1892, Ithaca, NY: Cornell University, December 1982.

14

Conclusions and Research Needs

RON ISAACSON
Gas Research Institute, Chicago, Illinois, USA

and

DAVID P. CHYNOWETH
Agricultural Engineering Department, University of Florida, Gainesville, Florida, USA

Anaerobic digestion, as a process for conversion of biomass and organic wastes to methane, has been demonstrated to be technically and economically feasible for a variety of feedstocks at a broad range of sizes from small-scale systems in individual household units to large-scale systems in wastewater treatment plants. Application at the commercial scale has been directed principally to digestion of domestic sludge and, to a lesser extent, livestock and industrial wastes. In applications of waste conversion, the principal objective has been waste decomposition, not necessarily methane production. The theory used for design and operation of these conventional digesters is based primarily upon empirical data and laboratory observations of the performance of the total fermentation and its response to variables such as loading rate, retention time, pH, alkalinity, and temperature. Little is known about the individual bacterial species or even the broad groups of bacteria involved in the fermentation. In fact, several of the important bacteria in the anaerobic digestion process have not yet been isolated. It is, therefore, not surprising that the process suffers from unpredictable performance with respect to conversion efficiencies, methane yields and production rates, nutrient requirements and stability. Although the degree of predictability can be improved on the basis of parametric laboratory experiments with specific biomass feeds, it should be recognized that each feed behaves differently. Even the most experienced researchers periodically encounter digester imbalance or failure for no understood reason; and recovery of imbalanced digesters is sometimes slow or even impossible. With the goal of producing pipeline-quality gas from biomass, the gas industry must view anaerobic digestion from a different perspective than currently prevails for waste or small-scale biomass conversion. Such digester systems may be larger than existing commercial units and receive feed quantities of 500–1000 dry tons or more per day. Feed availability may be variable and include several different biomass species. Because of the high cost of feedstock production and processing and the potential of large quantities of undigested residues with which to deal, conversion must be as complete as possible. In view of the large size of the digesters involved, the consequences of digester instability and the requirement to discharge the contents and restart the digester with a fresh inoculum could be monumental. Another major goal of biomass-to-energy digesters is to maximize net energy by optimizing the digester operating requirements. Finally, it is desirable that digesters be simple in construction and operation with low capital and operating costs. In view of these requirements related to large biomass-to-energy digester systems, the following major goals are recommended for research:

- Maximum conversion and associated methane yields.
- Maximum methane production rates.
- Maximum methane content of the gas.
- Minimum process energy requirements.
- Stable and predictable performance.
- Simplicity of design and operation.
- Minimum capital and operation costs.

While these goals are to maximize or minimize a particular parameter, the ultimate system developed will be an optimum one that has required trade-offs to be made. For example, minimum capital costs may fit well with maximum methane production rate, but the high cost of the feedstock may necessitate achieving high methane yields at the expense of methane production rate. This type of optimization must ultimately be done, but without knowledge of the limits of the system, i.e. maximum and minimum, it cannot be done effectively.

To best achieve these goals, a comprehensive, integrated program is recommended to study the anaerobic digestion of biomass. A limited number of representative feedstocks should serve as prototype feeds for a balanced, coordinated research effort including microbiology and biochemistry, microbial ecology, bench-scale process development and optimization, and experimental test unit studies to evaluate process scale-up and to define materials processing equipment. Each prototype species digestion program should be closely coordinated with studies on growth, harvest, and systems evaluation. It is anticipated that the resulting processes and associated data bases will be applicable to numerous species with properties similar to the prototype species. These test species might be separated on the basis of particulate solids content, biodegradability, and chemical composition, and should initially include marine, freshwater, woody, and grass species. In addition, the anaerobic biogasification potential assay should be employed for new, promising species as they become available from growth programs around the country.

The authors of the individual chapters and members of the Editorial and Advisory Panel (see Preface) compiled an extensive list of research needs for anaerobic digestion. The research needs were categorized in accordance with the chapter title and are listed in the Appendix. Each member of the panel then assigned a rank number (1 through 6) to each research need. In addition to assessing the importance of the need as high, medium or low, preference was given to long-term (>10 yr) over short-term (<10 yr) benefits. For example, high importance, long-term benefits would receive a 6, while low importance, short-term benefits would receive a 1. Research needs that obtained a ranking of 5 or 6 were deemed to be High Priority Research Needs (Table 14.1) that should be addressed first to facilitate the development of anaerobic digestion of biomass and wastes for the production of pipeline-quality methane.

The remainder of this chapter comprises a brief discussion of the major research areas: Microbial Ecology, Feed Characteristics, Process Design, Operation, Evaluation and Control, and Digester Development and Commercialization.

Microbial Ecology

With the exception of the rumen, knowledge of the microbial ecology of methanogenesis in anaerobic environments, including anaerobic digesters, is superficial. This can be attributed to the several factors:

(1) Isolation and study of anaerobic bacteria is time consuming and tedious.
(2) The fermentation involves numerous species.
(3) Anaerobic environments are different and involve different organisms.
(4) Interdependency of different species often restricts their ability to grow in pure culture.

TABLE 14.1
High Priority Research Needs

- An ecosystem approach to the study of key microbial activities (limitations and interactions)
- Factors limiting conversion yields of important bacterial isolates
- Interactions of isolates using co-culture and culture reconstruction techniques
- Standardization of the anaerobic biogasification potential assay
- Development of improved inocula through enrichment and acclimation techniques
- Effects of particulate solids on feed biodegradability, reactor configuration, and operation
- Structural properties of biomass and relationship to hydrolysis
- Mixed feed digestion and effects of changing feeds
- Nutrient requirements of hydrolytic, and transitional bacteria
- Effect of chelation on nutrient availability to microorganisms
- Methods for rapid digester recovery from unstable conditions
- Methods to promote and assure digester stability
- Automated analysis of digester performance parameters
- Develop improved digester designs and operating conditions for prototype biomass feedstocks
- Methods of improving microorganism retention, including attached film reactors, media, organism attachment, and sloughing
- Emphasis on net energy in planning and evaluation of results of research
- Development and evaluation of dynamic models for a variety of different digester designs
- Kinetics of non-steady-state conditions such as start-up, change in loading, change in feed
- Design and operation of experimental test units on location of biomass growth facilities to:
  - Scale up new digester concepts developed and tested at the laboratory scale
  - Integrate with growth, harvest, front-end and effluent processing
  - Obtain data needed for final process design and economic analysis
  - Develop new internals, such as influent distributors, gas separators and scum removers
- Develop gas cleanup systems specifically tailored to large-scale digester gas markets

The direct benefits of microbial ecology research to improve the design and performance of anaerobic digesters are limited. Rather, it adds to the pool of knowledge on the system, which ultimately will be applied to the understanding and improvement of the process. Therefore the microbiology and biochemistry of pure and mixed cultures should be investigated and the results related to observations on digester performance.

Elucidation of the microbiology of anaerobic digestion should approach the process as an ecosystem with identification of key microbial activities, their limitations and interactions.

Feed Characteristics

The physical and chemical characteristics of biomass feeds directly impact the methane yields and production rates from anaerobic digestion as well as the physical operation of the digester. No single digester design or set of operating conditions can be applied to all biomass species. A limited number of prototype biomass species can be selected on the basis of particulate solids content and chemical composition to serve as a basis for intensive investigation. The results of these studies will be translatable to other biomass species with similar composition. It should be emphasized that even within a single species, significant differences in chemical composition that impact digester performance can result, depending on growth and harvest conditions. The particulate solids content appears to have the greatest impact on selection of digester design. The anaerobic biodegradability is most significant in selection of candidate species for large biomass systems employing anaerobic digestion for conversion. Finally, the bacteria involved in the methane fermentation require a balance of nutrients for optimum conversion. These requirements are not well documented and the need for nutrient supplementation will vary with different biomass species.

Process Design, Operation, Evaluation and Control

Commercial application of the anaerobic digestion process is currently limited primarily to sewage sludge, industrial wastes, and animal wastes where the primary goal is waste disposal. The reactor types, operation, conditions and performance typical of systems operated on these wastes are not generally applicable at the scale that would be required for conversion of the large quantities of biomass envisioned for production of methane for supplementation of pipeline supplies. The design of such digester systems must be tailored to the availability and characteristics of the feedstock(s). They must be economic and capable of achieving high loadings with a high methane yield, low process energy requirements, and a high degree of process stability and reliability. To achieve this, a variety of innovative digester schemes should be developed and evaluated for several prototype feedstocks. These digesters should be able to achieve high solids and microorganism retention times and low hydraulic retention times. These reactor types will require development of sound theoretical and experimental bases for design, start-up, operation, evaluation, and control. Process models should be developed for prediction of their performance and should incorporate factors influencing net process energy. Translation of laboratory process development research into commercial systems should include a logical sequence of scale-up units that address scalability of the biological process as well as the numerous questions related to materials handling. These units would also provide data at a scale required for a valid systems and economic analysis.

Digester Development and Commercialization

Large-scale digesters receiving biomass from energy farms are not yet commercial. It is evident from this review that digesters employed for the production of gas in quantities of interest to the gas industry will differ significantly from conventional digesters in terms of feed rate, digester unit sizes, digester design, and method of operation. Laboratory studies in progress in the USA and other parts of the world have identified design and operating conditions for select biomass feedstocks that meet the requirements established for large biomass digesters: high yields and rates, process stability, and simplicity of design. This research should continue on several feeds to further improve design and optimization of the digestion processes. The next step is to develop anaerobic digestion research facilities where experimental test units can be operated in conjunction with growth, harvesting and front-end feed and effluent processing, to select the best equipment for materials processing, to evaluate selected reactor configurations for materials handling, and to assess scale-up performance. These facilities should be used to select the final process and obtain data for integrated process performance and economic assessment. Process economics must be conducted in conjunction with other elements of the total biomass growth and conversion scheme. Process economic models should be developed to facilitate this process.

Appendix: Research Needs and Prioritization

The authors of the individual chapters and the Editorial and Advisory Panel compiled a list of research needs for anaerobic digestion. These research needs are listed on the following pages.

The Editorial and Advisory Panel assigned a Rank Number for each need according to the following scale:

Rank	*Benefit time frame**	*Assigned rank number*
Low	Short-term	1
Low	Long-term	2
Medium	Short-term	3
Medium	Long-term	4
High	Short-term	5
High	Long-term	6

*Short-term, <10 years; long-term, >10 years.

The following is a listing of all the research needs that were compiled. They are grouped by Chapter. Individual rank numbers are shown as well as the average ranking. Research needs having an average ranking of 5 or 6 were considered as High Priority Research Needs and are discussed in Chapter 14.

Chapters 2–4: MICROBIAL BIOLOGY

Assigned ranking					
	Advisors			*Average*	
A	*B*	*C*	*D*	*ranking*	
					The microbiology and biochemistry of pure and mixed cultures isolated from prototype feed digesters should be investigated and the results related to observations on digester performance. The general approach for isolation and study of important organisms is presented below and should be done for the major groups of bacteria identified in this review: hydrolytic, transitional, and methanogenic.
6	4	5	5	5	Investigate anaerobic digestion as ecosystem with identification of key microbial activities and their limitations and interactions.
4	4	5	2	4	Isolate and characterize bacteria present in high dilutions and known to influence rate-limiting steps from digesters operated on different substrates and under different conditions (loading, retention time, temperature, etc.).
–	–	5	–	(5)	Determine longevity of various bacteria in a successfully operating digester.
4	4	5	3	4	Study activities of isolates present in high numbers and the relationship of activities to those observed in mixed culture.
2	5	5	5	4	Determine factors limiting conversion yields and rates of important isolates.
6	4	5	2	4	Determine physiological requirements, limits and establish conditions optimum for growth and metabolism.
6	5	3	5	5	Establish relationship of isolates with other groups of bacteria involved in the fermentation using co-culture and culture reconstruction techniques.
4	6	3	4	4	Elucidate important biochemical pathways and study enzymes and cofactors.
2	6	2	2	3	Employ genetic techniques for development and isolation of mutants with improved activity.

Chapter 5: FEED COMPOSITION

Assigned ranking					
Advisors				*Average*	
A	*B*	*C*	*D*	*ranking*	
2	4	2	5	3	Develop physical or chemical methods that can accurately predict the anaerobic biodegradability of all biomass feedstocks.
2	2	2	2	2	Develop a reliable technique to estimate the cell maintenance fraction for the digestion of particulate feeds in reactor types designed for cell retention.
3	5	5	5	5	Standardize the batch bioassay for determining ultimate biodegradability of feedstocks suitable for anaerobic digestion.
2	6	3	3	4	Study the effects of biomass growth conditions on biomass composition and the effect of composition on theoretical methane yields and biodegradability.
5	4	2	4	4	Identify and quantify the biodegradable and refractory components of biomass feeds and develop a relationship between chemical composition and biodegradabilty for all biomass feedstocks.
3	6	4	5	5	Study microbial inocula from various sources using enrichment and acclimation techniques for potential degradation of refractory compounds or to increase methane production rates of biodegradable compounds.
5	4	5	5	5	Study the effects of particulate solids on feed biodegradability, reactor configuration, and operation.
6	5	6	5	6	Study structural properties of biomass and relationship to hydrolysis.
3	5	3	3	4	Investigate mixed feed digestion and effects of changing feeds.

Chapter 6: PRE-/POSTTREATMENT

Assigned ranking					
Advisors				*Average*	
A	*B*	*C*	*D*	*ranking*	
6	4	3	3	4	Search for potential pretreatment methods that may not necessarily enhance glucose production, but some other intermediates, so long as these intermediates can be made efficiently from cellulosics and can be converted efficiently into methane.
4	5	2	1	3	Examine the potential for more sophisticated biomass processing schemes with integrated production of multiple products such as other fuels, chemicals, animal feeds.
2	3	3	1	2	Examine alkali pretreatment on different biomass feeds to evaluate its effectiveness toward increasing methane production rates and methane yields.
2	3	3	2	3	Investigate particle size reduction as a means of increasing methane yields and methane production rates from biomass feedstocks.
2	2	2	4	3	Study decarboxylation of carbohydrates as a general process and pretreatment scheme for increasing the methane yield.
3	4	4	3	4	Evaluate lignocellulolytic fungi for pre-/posttreatment.
3	5	3	5	4	Evaluate production of valuable coproducts by pre-/posttreatment.

Chapter 7: NUTRIENT REQUIREMENTS

Assigned ranking					
	Advisors			*Average*	
A	*B*	*C*	*D*	*ranking*	
4	6	5	5	5	Study nutrient requirements of hydrolytic, intermediate, and acetogenic bacteria.
5	5	5	3	5	Study effect of chelation on nutrient availability for microorganisms.
5	3	5	2	4	Determine requirements of key nutrients such as N, P, S, Fe, Ni, and Co as a function of reactor type and operating conditions.
2	3	5	4	4	Extrapolate nutrient requirements derived from batch growth systems to CSTR and other reactor types.
3	3	5	4	4	Develop a model based on chemical costs of each nutrient for process optimization of the rate and amount of methane production.

Chapter 8: TOXICITY

Assigned ranking					
	Advisors			*Average*	
A	*B*	*C*	*D*	*ranking*	
2	4	5	3	4	Determine the potential of the consortium of bacteria involved in anaerobic digestion to accommodate chronic or sporadic toxicity, the relation between type of toxicant, concentration, and duration of exposure.
2	5	3	5	4	Classify toxicants in anaerobic digestion.
2	5	5	4	4	Evaluate differences between unacclimated and acclimated toxicity.
3	1	5	4	3	Determine the effects of biomass pretreatment products, pesticide residues, and extractives from softwoods and hardwoods on the methanogenic fermentation process.
3	2	5	5	4	Clarify the reversibility of toxicity as a function of digester configuration.
2	2	3	3	3	Conduct a more comprehensive viability assay for the microorganisms responsible for key steps of anaerobic digestion.

Chapter 9: REACTOR DESIGNS FOR BIOMASS DIGESTION

Assigned ranking

Advisors A	B	C	D	*Average ranking*	
5	5	5	5	5	Identify design and operating conditions for select biomass feedstocks that meet the requirements established for large biomass digesters: high yields and rates, process stability, and simplicity of design.
5	6	5	4	5	Continue development of 'novel' digester concepts capable of improving yields and production rates for biomass feedstocks.
–	–	–	–	–	Investigate innovative methods of digesting feeds with suspended solids in the range of 3–10%.
2	2	5	5	4	Determine effects of solids residence time on digester performance.
2	2	5	5	4	Determine effect of SRT on nutrient requirements and cell synthesis.
3	3	6	5	5	Study methods of promoting long microorganism retention, in attached film reactors. These should include media types, factors influencing attachment and sloughing, and importance of attached versus interstitial organisms.
3	2	5	5	4	Clarify relative significance of solid, microorganism, and hydraulic retention time on digester performance.
3	6	2	5	4	Direct research on digester design and operation specific to physical and chemical composition of biomass.

Chapter 10: START-UP, OPERATION, STABILITY, AND CONTROL

Assigned ranking

Advisors A	B	C	D	*Average ranking*	
3	4	3	4	4	Develop and evaluate inocula derived from natural environments receiving substrates resembling test feed.

3	4	3	5	4	Start-up digesters, in particular attached film digesters, and determine effects of inoculum, medium, feedstock, and operating conditions.
6	6	3	5	5	Develop methods for rapid digester recovery from unstable conditions.
2	5	1	3	3	Develop accurate method for determination of alkalinity in presence of suspended solids.
5	5	1	3	4	Develop techniques for reducing volatile acids concentrations, such as dialysis or two-phase digestion.
2	2	1	4	2	Document ORP as a method of monitoring digester instability. Learn to control ORP for restoration of stability.
3	5	1	3	3	Determine the effect of hydrogen or other parameters as measures of digester stability. Determine the turnover rate constant of hydrogen.
3	3	–	2	3	Emphasize start-up, operation, and maintenance of stability in a thermophilic digester.
2	2	1	3	2	Investigate the effects of pressure (positive and negative). Use fermentation to raise pressure to pipeline levels.
3	4	4	4	4	Study the effects of mixing on activities of overall and specific populations.
3	4	3	5	4	Study the effects of mixing on mass transfer and materials handling of biomass.
6	6	4	5	5	Develop methods to promote and assure stability in digestion. Develop methods of recovering failing digesters.
5	6	3	5	5	Automate monitoring of performance parameters such as volatile acids, methane content of gas, hydrogen, pH, alkalinity, and ORP. Study continuous change in these parameters under conditions of continuous feed and various states of performance.
5	4	2	5	4	Automate feedback control of digester operation.

Chapter 11: KINETICS AND MODELING

Assigned ranking					
	Advisors			*Average*	
A	*B*	*C*	*D*	*ranking*	
3	4	4	4	4	Develop unified theories with basic kinetic equations for steady-state digester conditions, including start-up, and verify these experimentally.
5	6	4	5	5	Develop dynamic models for the variety of digester designs being investigated. They should predict the methane yield and production rate, methane content of gas, volatile acids, pH, and alkalinity as a function of operational parameters and performance.
3	6	4	5	5	Develop the relationship between kinetic parameters obtained during steady-state and non-steady-state conditions caused by loading shock and/or change in the feed.
4	4	4	4	4	Refine kinetic models to describe the response of digesters to toxic substances. These should include inhibition, adaptation, and recovery.
2	5	6	2	4	Include effects of unattached (or interstitial) microorganisms in attached-film (or fixed-film) reactors in kinetic analysis for developing a model.
2	2	3	5	3	Include endogenous metabolism, which is different at different stages of anaerobic digestion, in the kinetic equations for each state (for example, start-up, pre-steady-state, and steady-state conditions).
5	5	2	4	4	Develop hydrolysis models for various biopolymers (in pure and mixed forms).
3	3	4	1	3	Develop kinetic models for requirements for macro- and micronutrients.

Chapter 12: NET ENERGY

Assigned ranking: Advisors A	B	C	D	Average ranking	
2	6	5	6	5	Emphasize net energy analyses in digestion studies.
2	4	5	2	3	Evaluate the effects of the digester size on digester heat loss.
3	3	–	2	3	Calculate pumping energy requirements for some of the unconventional digesters such as fluidized- and expanded-bed digesters, with regard to bed height, degree of fluidization or expansion, and the physical properties of the material used for the bed.
3	4	1	5	4	Determine the effects of rheological properties of different feedstocks on energy needs for pumping, mixing, etc.
5	4	–	3	4	Evaluate mixing energy requirements for lignocellulosic materials with regard to their fluid properties.
3	5	1	5	4	Determine the effects of heat of reaction on digester heating requirements of feedstock, particularly with feeds having high solids content.
2	5	6	5	5	Consider use of heated water and heat exchangers between effluent and influent streams to conserve energy in net energy analyses.
2	6	6	5	5	Consider potential energy impact in establishing research priorities.

Chapter 13: DIGESTER DEVELOPMENT AND COMMERCIALIZATION ECONOMICS

Assigned ranking					
Advisors				*Average ranking*	
A	*B*	*C*	*D*		
3	5	6	5	5	Document the effects of scale-up during the development of various digester schemes.
3	6	5	5	5	Establish anaerobic digestion research facilities where experimental test units can be operated in conjunction with growth, harvesting, front-end feed, and effluent processing to select the best equipment for materials processing and to evaluate selected reactor configurations for materials handling and scale-up for performance.
4	4	5	5	5	Use ETU facilities to select the final process and obtain data for integrated process performance/economic assessment.
–	5	4	5	5	Initiate system studies to develop improved process flow concepts for planting, harvesting, storage, transportation, size reduction, pretreatment, digestion, gas clean-up, effluent treatment, and energy conservation for appropriately sized biomass conversion plants.
3	5	4	6	5	Develop improved digester internals (such as influent distributors, media, gas separators, and scum removers) tailored to biomass feedstocks.
2	3	6	6	4	Develop improved, low-energy mixing systems.
3	3	2	6	4	Develop automated digester instruments and controls to improve the stability of operation and the continuity of gas production rates.
3	5	6	6	5	Develop gas clean-up systems specifically tailored to large-scale digestion gas markets.
5	5	5	6	5	Widen the available data base for determining the sensitivity of gas costs to biomass feedstock costs, harvesting costs, pretreatment costs, digester costs, gas clean-up costs, and other pertinent cost elements.

5	5	5	6	5	Develop an appropriate gas cost formula for gas production from digestion of biomass feedstocks. The cost formula should be applicable to all digester types and cover a broad range of biomass feedstocks. For comparison purposes, a common basis should be defined for the project and financial elements of the formula.
5	4	–	1	3	Develop a generic economic model for digester gas cost.

Index

Printed in the United States
72507LV00001B/2

9 781851 660698